Curvology

Also by this author

Making Babies: The Science of Pregnancy

The X in Sex: How the X Chromosome Controls our Lives

Beyond the Zonules of Zinn:
A Fantastic Journey through your Brain

Teenagers: A Natural History

Middle Age: A Natural History

The Origins and Power of Female Body Shape

Curvology

DAVID BAINBRIDGE

Portobello
BOOKS

Published by Portobello Books 2015

Portobello Books
12 Addison Avenue
London
W11 4QR

A CIP catalogue record for this book is available from the
British Library.

1 3 5 7 9 10 8 6 4 2

ISBN 978 1 84627 550 0
eISBN 978 1 84627 551 7

Text designed by Lindsay Nash

Typeset in Minion by Avon DataSet Ltd, Bidford on Avon,
Warwickshire

Printed and bound by CPI Group (UK) Ltd, Croydon CR0 4YY

www.portobellobooks.com

Contents

I paint a woman's big rounded buttocks so that I want to reach out and stroke the dimpled flesh.

Peter Paul Rubens

You can never be too rich or too thin.

Wallis Simpson

From a visit to the newsstand . . .

'All working mums would be a size 8 [US 4] like me
if they weren't so lazy.'

'Kate shows off her new curves but is slated for her alcohol tummy.'

'Waist of an eight-year-old!'

'Report: The women having surgery for the school run.'

'Adele – every inch a star!'

'I feel so humiliated – I look six months pregnant.'

'You in 2015 – slimmer, happier, stronger.'

'The insider's guide to the best quick fixes.'

'I'm sexier with the wobbly bits and cellulite.'

An introduction to our obsession

If God hated flesh, why did He make so much of it?.

Stranger in a Strange Land, Robert Heinlein, 1961

'It was just that feeling of not being good enough – desperate to be liked. I thought if I was thin then I'd be okay – I'd be liked.'

Anonymous interviewee 'D'

This is a book about the female body – its biology, the mind it contains, the culture that surrounds it – and why it has turned out to be the strangest thing in existence. It has its origins, it has its future, and it also has immense power right *now*. Its influence pervades our lives from our base visceral functions to the esoteric artefacts of our modern civilisation. And it dictates what it feels like to be a human being – female or male.

Modern Western society is obsessed by the female body. Magazines lure readers of both sexes with it; the fashion industry accentuates, conceals or distorts it; we are continually told that it is becoming too large, too small, too exploited. We rehash questionable old statistics about how much time women spend looking in mirrors, yet ridicule women who go on 'mirror diets' – deliberately avoiding their own reflection. Women complain about the pressure they feel to conform to an 'ideal' body shape. We fear that young women are being psychologically damaged by the barrage of perfect, often heavily airbrushed female flesh they see, yet we persist in consuming the products of the media who proffer that flesh. The female body is, quite simply, everywhere.

Yet rarely do women seem content with their bodies. Research shows that between 50 and 80 per cent of women are dissatisfied with their bodies, a consistently higher value than for men. At any one time, far more women than men describe themselves as being on a diet. Fifty per cent of women say that they would change the shape or size of their breasts if they could. Parents, friends, enemies, the media, even dolls – all have been shown to affect women's self-image.

People have higher expectations of what attractive women can achieve – and studies show that we assume skinnier women are richer and better educated. Even in the cradle, female babies possessing (objectively measurable) 'cute' characteristics receive more parental attention. Appearance and attractiveness have been described as the 'personal billboard' we present to the world, and creating that female fleshy billboard can be an unforgiving process. Women's bodies vary enormously, and every day we all deal with that variation.

There were two simple observations which spurred me to study female body shape, and they will reappear in different forms throughout this book.

The first observation is that we are the only species in existence with curvy females, and this bizarre uniqueness demands an explanation. We are happy for animals to be considered 'handsome' or 'cute', but the idea of curvaceous she-monkeys, voluptuous mares or buxom sows strikes us as distasteful. We resolutely reserve the concept of curvy femininity for humans, and I believe that our uneasy reaction to breaking that convention demonstrates how we all instinctively understand the uniqueness of human female body shape. As will become clear, this exceptional female morphology has dramatic effects on women's lives – it affects how they move, the diseases they suffer, their sense of self, and how other people treat them. And sometimes those effects can seem counterproductive – for example, how did our species evolve to a point at which most females cannot run naked and unsupported without experiencing actual physical pain?

My second observation is that women think about their bodies more, and in fundamentally more complex ways, than men think about theirs. Biologically, it makes sense for individuals to want to look good to the opposite sex, because this is how you attract a high quality mate to help you produce strong, successful offspring. However, I do not think that heterosexual women's cogitations about their own bodies are often a response to that biological drive. Women think about their bodies a lot, and to a level of complexity and subtlety which amazes most men, but I would argue that they do not usually have men in mind when they do it. Indeed, I suspect that this is why they may often have misconceptions about what female bodies heterosexual men like, and how 'adaptable' men can be if the fleshy reality differs from the theoretical ideal. Compare, for example, the women's bodies on the covers of women's and men's magazines: which vary more? As we will see, men do feature in the story of the female body, but not as much as they might hope – and there are potent evolutionary, sexual, psychological, social and cultural reasons why.

My intention in this book is to discover where all this came from – all the fascination, debate, unease and even fear about women's bodies. It may seem obvious why men think about female body shape so much, but it is not immediately clear why women think about it even more.

I am a reproductive biologist and a vet, I have a zoology degree, and I teach anatomy and reproductive biology to university students. I have written books on pregnancy, sexuality, the brain, teenagers and middle age, and I am a forty-something Caucasian male. I am not sure if this makes me the ideal person to write a book about human female body shape, but I am also not certain who the ideal person would be. I could be argued to be a dispassionate observer or biased voyeur, depending on one's point of view. One thing I do know is that men think about women's body shapes a great deal, but I also know that they find women's ways of thinking about those bodies bewildering.

Humans are the most unusual and complex animals on earth, so there can be no simple answer to why female body shape is so important and contentious. By standing back from the everyday mêlée of arguments about women's bodies, and taking a more 'zoological' approach to this uniquely human phenomenon, I believe I can answer some questions which others do not like to ask. After all, a purely cultural or sociological approach cannot explain the power of the female form without an underpinning of evolution, biology and psychology. Each of us has inherited a central, non-negotiable, biological core of what female bodies are and how we react to them. Culture and society are extremely important of course, but they are changeable things which flutter impermanently around that biological core of femaleness. So, any cultural or social investigation of women's bodies which did not build on basic biology would be without a genuine context, rootless.

Taking an evolutionary and zoological approach to the origins and power of female bodies will lead me to focus primarily on heterosexual relationships. It is true that homosexual women constitute a minority of the human population, and little research has focused on the role of the female body in same-sex relationships, but there is another reason why I must focus more on heterosexuality. This is the simple fact that the story of the evolution of our species is entirely dominated by heterosexual, child-producing unions. If you think about it, *all* of our ancestors fall into that category – it is only *their* genes which have survived to the present day. Heterosexual humans are the ones on whom the vicissitudes of natural selection have acted to produce the human race as we see it today.

Because of the multifaceted nature of our relationship with women's bodies, I have divided this book into three parts – body, mind and society. The first is focused squarely on the bodily biology of womanliness – how human women acquired their unusual shape, how each young woman now acquires it anew, the dramatic variations in women's body shapes and the surprising effects they have on health, physicality and fertility. In the second part, I will look at how women's bodies affect the human mind – what it means to inhabit one, what it is like to desire them, and how modern women negotiate the conflicting pressures of food, mood and shape, sometimes successfully, sometimes not. And, in the third part, I will investigate the relationship between women's bodies and the world outside – how different cultures and social environments judge, modify, conceal, celebrate and condemn the female body. Only then, with this combined physical-mental-societal understanding of women's bodies will it be possible to finally explain, in the last chapter, why they obsess us so.

Women's bodies are far more womanly than they absolutely need to be. All that most other animals need to function as a female are some discreet internal plumbing and inconspicuous mammary glands, but something much more radical has happened to human females. And for the first time in our history we now understand bodies, minds and societies sufficiently to comprehend exactly why women's body shapes affect us the way they do.

At last we can piece together the story of how something as complex as women's attitudes to their own bodies – how they affect relationships with men and other women, and even how we raise our daughters and sons – have been forged during millennia of evolution in the cauldron of human sexuality, thought and society, and a sheer desperation to survive.

PART I

THE BODY

'In the morning I wake up and I think, "Ooh, you look good – you look skinny," but by the end of the day when I've eaten all that food I think, "Oh God, you're really fat. Lose some weight." I did go on a diet a few weeks ago and I did lose five pounds. No – three pounds.'

Anonymous interviewee 'A' (age 21, body mass index 22.7)

'When I first started developing I was absolutely terrified. I remember wanting to remain childlike, and everything that meant. I felt like there was a loss of innocence that I should be ashamed of. There was a phase of a few months when I was about thirteen when I just stopped eating altogether. When I was sixteen I suddenly woke up with huge boobs and I was so embarrassed about them. I didn't know what to do with them – I just felt apologetic about my body . . . I love them now.'

Anonymous interviewee 'B' (age 32, body mass index 26.1)

'All the time – I'm always thinking about my body. If I'm due to go out, in the week leading up to it I'll be panicking the whole time about looking right. It constantly preys on my mind.'

Anonymous interviewee 'C' (age 33, body mass index 21.3)

'I would never walk around naked in front of someone. It's my backside – he'd dump me in a second. I think my backside looks revolting so I assume men would think the same . . . I don't want to spend the rest of my life feeling like this.'

Anonymous interviewee 'D' (age 40, body mass index 24.5)

'I used to think I was fat, although I now realise that I'm not fat and never was fat, but I used to be careful not to show too many squashy bits. I might have been breathing in to stop my stomach looking fat when it wasn't. I got it wrong.'

Anonymous interviewee 'E' (age 70, body mass index 24.8)

(Body mass index is a frequently used assessment of body weight, calculated as body weight in kilograms divided by the square of the height in metres. Figures between 18.5 and 25 are sometimes said to lie within the optimal range.)

ONE

Where women's bodies came from

What is it about these preternaturally small women?

Last Night in Twisted River, John Irving, 2009

'I can't really run at full pelt – I think I would knock myself out – just because of the boobs – they'd go everywhere.'

Anonymous interviewee 'B'

In the rust-red *light of another dawn the girl gazed down at her thighs. She pinched them – they were still smooth and full, and she felt them sway slightly whenever she walked. She was not sure, but she did not remember them always being like that.*

Six dry seasons ago her mother had talked to her about this. She told the girl that she had already lived for twelve dry seasons before she was traded. The girl did not clearly remember being traded, but she remembered many people from a strange tribe – women and men – all looking at her closely. Especially, they had looked at her legs and her bottom. She remembered being told her thighs were full and strong like her mother's, and this was why she would be cherished by her new tribe and her man. He was the son of the leader of her new tribe and she had wondered why her old tribe – the tribe of her childhood – was giving her to him. One of her friends was traded at the same time, but she was thin and the bones in her bottom stuck out. Her thighs had never been smooth and she had died two dry seasons ago – she could not grow the baby inside her, they said.

The girl sat up and looked at the sleeping women around her. They still felt like strangers even after six dry seasons. She missed her mother.

    The men were away, scaring off the beasts. Times had been harder recently, so the old women said, and the beasts now ventured closer than ever before. They were faster and angrier than people, but the men had learnt that they could scare them away if they stood tall and threw their sharpened stones at them. Even lions were scared of the upright people and their stones. Some of the men now wanted to go out and find animals on purpose so they could kill them and eat them. They were tired of scavenging, they said, and if they had to scare the animals away, they might as well kill some for food. Everyone argued about this, about whether it was safe, but everyone knew that the children needed lots of food to grow up strong, and that there were not as many plants to eat any more. The girl's stomach was empty, and the emptiness gnawed at her.

    The old people – the ones who had seen sixty or seventy dry seasons – always talked about how the rains used to be much stronger when they were young. Now there were dead tree stumps everywhere, but even the girl remembered walking through mud and guzzling berries when she was little. There was much less food now, and the children would have to be cleverer to find enough to eat when they grew up. Why had the rains faded away? Why did they no longer have trees to shelter under? Why should only the smooth-thighed girls have babies now? And the girl wondered why her new tribe always had to move around – never resting in one place for long. She had not bled for a while now, and deep in her belly she could feel something wriggling all the time. She clasped her hands over it. She feared for it.

Humans are strange animals, and they evolved in strange times. Very little remains of our ancestors – the few crumbling fragments of tooth and bone we pick from the ground are not enough for us to work out the genealogy of the human lineages which meandered away from the split from our closest relatives, the chimpanzees, to the present day. The names are evocative – *Sahelopithecus, Ardipithecus, Australopithecus, Homo* – but they conceal a jumbled, tangled family tree, or rather a family thicket, which we have not yet penetrated.

Of course some fossil people are men, and some are women, but sometimes we cannot tell. Often the fossil people are so fragmentary that they remain sexless.

Yet it is clear that, over the last eight million years or so, hominins gradually acquired an array of exceptional characteristics – features that set us apart from the beasts. This was an erratic process, and often several ancient proto-human species coexisted, each with its own admixture of ancestral and modern human traits. Many of the great transitions in human evolution probably occurred in response to terrible changes in our ancestors' environment – increased aridity, especially, may have forced hominins to adapt to new ways of life, away from the forest, and eventually away from the trees altogether.

It is hard to imagine just how traumatic these transitions must have been. Climates can change remarkably quickly, and one can only imagine the desperation with which our forebears must have struggled to cope with these upheavals. In this light, the dramatic re-engineering of the human body can be seen not so much as a triumph of human guile and adaptability, but rather a sad, scrabbling litany of pain and suffering which contorted us into our present, bizarre form. Whether it is the way we walk, the way we think or the way we bear children, our bodies changed out of all recognition, and sometimes that process was frighteningly fast.

The original change which launched the human story was probably not related to brains or babies. Instead, walking upright on two legs got us started. Other primates may occasionally waddle around on their hindlimbs for short periods, but none of them walks the way we humans do. We retained the cylindrical limbs, powerful buttocks and grasping hands of our primate ancestors, but everything else was redesigned when we became fully bipedal.

Anthropologists used to assume that humans started walking on two legs to do noble things like holding tools, or even carrying babies, but now they are not so sure. Most primates rear up on their back legs to threaten or intimidate, and ancient humans may have stood erect for precisely the same reason. Indeed, no matter how weak, slow and puny we upright humans may be, wild animals seem to find us strangely intimidating. This need to stand up and fight may also have driven the evolution of another distinctive human feature, as recent studies suggest that our hands developed their distinctive shape not to manipulate the artefacts of a developing civilisation, but to form fists.

Humans walk in a wonderfully unusual way, bearing their weight vertically aloft and propelling themselves on one leg while the other swings effortlessly forward before striking the ground. Our skeletons are so well adapted to this unique mode of propulsion that surprisingly little muscular effort is required to maintain it. Vertebrae in our lower spines are wedge-shaped, which means that our lower backbone bows forwards, throwing the weight of our heads, chests and abdomens over our hip joints, where it can be carried more efficiently. Our legs are long and straight and our arms short and slim to bring our centre of gravity down towards our hips. Our leg bones have been honed over the millennia so that when we walk, our bodies make almost no wasteful up-and-down or side-to-side movements. Our foot has lost its grasping ability, has gained its arch, and its skeleton now resembles that of a slender bear rather than any monkey. And of course our eyes must still gaze horizontally, despite our upright stance, so our brain and skull have developed a novel ninety-degree kink so that we do not spend our lives looking vertically up at the sky.

The one bone which changed the most, and which contributed most to human bipedalism, was the pelvis. This comes as little surprise because the pelvis connects the legs to the trunk, yet more intriguing is the effect the reshaping of the pelvis has had on the outline of humans' bodies. The great apes' pelvises are long and straight, to drive propulsive forces from their legs along their horizontally oriented spines. In contrast, even by the time of *Australopithecus* – around four million years ago – the hominin pelvis was already kinking, to draw the legs under a more vertical body. Also, the pelvis was becoming shorter and squatter, to bring the centre of gravity downwards to a stable position near the hips. The kinking of the pelvis threw the buttocks out backwards, producing the prominent rounded human bottom, while the shortening of the pelvis created a long boneless gap of narrow abdomen between pelvis and ribs – a waist, in other words. So all those millions of years ago, humans already had waists and bums, and we will see later why those two novel features became so emphasised in the female of the species.

I teach animal anatomy, and to me a chimp skeleton looks like an oddly shaped quadruped – a jackal or a boar – yet the human skeleton looks nothing like that at all. As far as we can tell, our array of adaptations for bipedalism seems to have been present from very early in the human evolutionary story. Straight, cylindrical limbs rooted in compact globose buttocks and shoulders; a floating effortless stride and a high, forward-facing head – *Homo*'s body existed long before *Homo*'s mind.

The phrase 'early man' reflects a chauvinism which is gradually being purged from palaeoanthropology. It may seem obvious to state it, but men and women have always evolved together, in parallel, although you might not believe it if you read some older anthropology texts. However, despite this shared story, as far back in our history as we can tell, human women and human men's bodies have been exuberantly different.

Humans are 'sexually dimorphic': the two sexes differ, on average, from each other in certain characteristics. This phenomenon is not unique to humans – in fact many species are dimorphic, and often much more dimorphic than we are, yet few of them are dimorphic in as many intriguing ways as human females and males. Indeed, women's bodies differ from men's far more than is strictly necessary to gestate, bear and nourish children, and many of these differences seem to have arisen during the several million years since we split from the chimps.

The most obvious human dimorphism is size. Men tend to be bigger than women – perhaps 1.07 times taller and 1.15 times heavier, although these figures vary between populations, and presumably also depend on diet and health. However, this degree of size dimorphism is not actually very great – chimps have a weight dimorphism of approximately 1.30, for example – and human sex disparities seem to have ebbed and flowed over the course of evolution, with *Australopithecus* possibly having a height dimorphism as great as 1.50.

So rather than worrying about why men are larger than women, we should perhaps be wondering why human size dimorphism is so *slight*, especially today. In many animals size dimorphism is a sign of polygamous or promiscuous breeding systems and between-male aggression. The theory is that males who are large can exclude smaller males from mating with females, and thus perpetuate their big-male genes. This suggests that the relatively small differences in size between the human sexes reflects an intriguingly 'nearly-monogamous' natural pattern of breeding in our species. However, there could be other explanations for our limited dimorphism as well.

For example, over the course of evolution, both human sexes have become much larger, perhaps doubling in size, and this hints at other forces which may be at work on us. For a start, larger animals can be more energy efficient than small animals, and they are certainly better at scaring away predators. However, a further, unique change has taken place in the human way of living – our brains have grown out of all proportion to everything else. One of the changes that allowed this brain expansion was that women started to give birth to babies with unusually large heads – and throughout human history larger women have been more likely to survive this challenge. Thus one of the reasons that humans became larger, and especially human females, was so they could successfully bear large-brained children. So there are compelling reasons why in humans in particular, females are not much smaller than males. If men got any bigger or women got any smaller, obstetric disaster would have expunged us from the face of the earth.

But of course women are not just small men, so disparities in size are not evenly distributed throughout the male and female frames. Most things are absolutely bigger in men, but not all. The male-divided-by-female ratio for arm girth may be 1.06, or for waist girth 1.05, but the ratios for hip and thigh girth are both somewhere near 0.96 – average hips and thighs are *absolutely* smaller in men than in women. Thus the sex differences in body dimensions are complex, and demonstrate the obvious fact that the two sexes differ not only in size, but also in *shape*.

As it turns out, almost every part of the human skeleton is affected by sex differences. Indeed, the all-pervasiveness of these differences is striking: women and men really do seem to differ far more than they have to. Particularly striking is the fact that it is usually women who possess the most exaggerated version of the features which set modern humans apart from our ancestors, whereas men often exhibit older, ancestral traits (overall size is one exception to this). We do not know why women should be more 'modern-looking' than men, but it has been suggested that many of the characteristics men find attractive about women are the same characteristics which make them look distinctively human. So perhaps these preferences are a relic of a time when there were several hominin species intermingling on the plains, and *Homo* men had to be sure to mate only with *Homo* women.

For example, the human face has changed a great deal over the last several million years, and some of those changes are particularly emphasised in women. Among other changes, humans seem to have lost many of their ancestors' adaptations for chewing – our 'muzzle' is short and our face flat, and especially so in females. Humans generally, and women in particular, have small teeth and notably weedy canine teeth. Many primates have prominent bony ridges on their face, and in their brow especially, probably to buttress against the forces exerted by chewing muscles. However, these struts have dwindled in humans, almost disappearing in women. We are not sure why chewing became less important in ancient humans – maybe they ate softer foods, or more likely they rendered them less chewy by cooking – but the legacy of this change remains in the human, and especially the female, face.

The enormous enlargement of the brain has led to our species having a distinctively domed head, as well as an unusual skull-to-spine connection which allows that big head to balance atop the body.

Once again, women seem more 'modern' in this respect, and their skulls appear more domed than men's, although the reasons for this are probably complex. Women's brains are *relatively* larger than men's, but they also look more cerebral because they have larger foreheads unencumbered by masculine brow-ridges, and smaller chins which enhance a 'top-heavy' facial appearance. It has been suggested that the distinctively domed human cranium, and indeed the feminine cranium shape, evolved by retention into adulthood of the domed heads and forward-facing eyes of our ancestors' infants – and indeed, baby chimps and gorillas still look more like humans than adult apes do. According to this theory, all the skull had to do to look human was simply to never grow up.

Another manifestation of bodily dimorphism is the fact that those remarkable sauntering human limbs come in two very different versions. An unusual angulation of the leg bones means that women's legs make more side-to-side movements, although this famous 'wiggle' probably has only a small detrimental effect on the efficiency of walking or running. Women's feet are relatively smaller than men's lumpen appendages and their sole arches are more pronounced, a difference already present by the age of twelve months, and there are many other differences in the bones of the foot.

Women's hands are relatively smaller too, and their arms shorter. As well as many other differences in the small bones of the hand, the ring finger is usually shorter than the index finger, the result of low male sex hormone concentrations during development. Another strange difference between the sexes lies in the angulation of the elbow, or 'carrying angle'. If a man extends his arm, palm uppermost, he will see that his forearm lies in near-alignment with his upper arm, whereas in women the forearm tends to bow out to the side. It is not easy to measure the carrying angle, but studies suggest that after puberty the angle of deviation may be less than ten degrees in men, compared with fifteen degrees in women. This pronounced out-bowing of women's arms is difficult to explain, although it may allow their arms to swing from their narrow shoulders without striking their broad hips. However, even this theory cannot explain why the carrying angle is usually more pronounced in the right arm of right-handed women and in the left arm of left-handed women.

In general, men's bones are larger and more robust than women's. They deposit extra bone around the outside of their long bones which makes them stronger, whereas women lay down bone inside their long bones which contributes little to strength, but may act as a store of calcium, perhaps for milk production. These differences probably explain why women suffer more fractures as they age – both sexes lose bone-mineral, but men's bones have the advantage of being much stronger to start with. Women's joints are also disproportionately small and weak, and they carry less muscle mass than men. Following exercise, women's muscles enlarge less than men's, and strength differences between the two sexes can be dramatic – men's hand grip is on average twice as strong as women's.

Finally, women's and men's torsos show some of the most dramatic sexual dimorphisms of all. The thorax, especially, is surprisingly different – ribs, vertebrae, breastbones and collarbones all exhibit differences between women and men. The main result is that the chest is much smaller in women, and not only does this decrease women's ability to extract oxygen in their lungs, but it also transforms their appearance. A small thorax means their shoulders are not as broad, enhancing a 'bottom-heavy' appearance, and it also makes the female neck look longer and more sinuous than the male. The small ribcage has also permitted an elongation of the abdominal cavity. From breastbone to pelvis, women's abdomens are relatively longer than men's, providing more space for a growing baby. As human babies evolved to become larger, more space was needed to house them, until we are left with the situation today in which the pregnant uterus at its greatest extent presses against the diaphragm, compressing women's lungs and even nudging their heart over to one side.

So even before we examine hips, fat or breasts, the bodily differences between the two sexes are already profound. If, in our far distant future, anthropologists from an advanced civilisation dig up the bones of our own species, they could be entirely ignorant of women's curves and still marvel at the evolutionary forces which drove our two sexes to be so skeletally different.

The pelvis – eight bones which unite to make one big one – played a central role in the evolution of human female body shape. Indeed, in many ways the pelvis was the armature about which the architecture of the female body was constructed. Already distorted by the demands of bipedalism, the pelvis then went through a second, female-only stage of development.

The internal size of the maternal pelvis is the single greatest factor limiting a baby's chances of being born successfully, so females of almost all mammalian species have wider pelvises than males, even if those males are otherwise much larger. In most mammals, the widest part of a baby is its shoulders and chest, but in some primates the brain has become so large that the head now presents the greatest challenge. For this reason, getting the head through that pelvis is a matter of life and death for female primates, which is why there is a clear trend for big-brained primate species to have wider female pelvises.

Beset by terrible climatic disruptions, it seems ancient humans coped mainly by becoming cleverer. Their brains became bizarrely large, giving them lots of extra thinking-capacity so they could work out how to survive in their ever-deteriorating environment. This was an unimaginably difficult time, when spearing a fish or digging up a tuber could make the difference between your children living or dying, so the pressure to get smart was intense. And it is because of this pressure that the human brain has ended up being perhaps five times bigger than it should be for an animal of our size.

Yet this valuable new brain – this oversized organ which apparently made the difference between extinction and survival of our species – was fraught with problems. It consumed alarming amounts of energy and it took a painfully long time to develop, but most of all it endangered expectant mothers. By this time the hominin pelvis had become short and slim to allow efficient bipedal walking, but now this had to change, in women at least. The female pelvis was now forced to widen again to allow new, larger baby heads through. A conflict arose between the need to walk and the need to give birth, and an uneasy compromise was reached.

First, the fetal brain played its part by delaying its own growth – keeping as small as possible at the time of birth. Despite its eventual huge size, most of the dramatic enlargement of the human brain is delayed until *after* birth. In most newborn animals brain growth slows abruptly, but uniquely in humans the brain does most of its growth postnatally. In many ways, a human infant's brain grows as if it were still a fetus – adding one million new brain cells every twenty seconds in newborn babies, and then enlarging with spectacular rapidity until the age of six or seven.

However, there seem to be limits to how much the brain can postpone its growth until after birth, and the fossil record also recounts how the pelvis met its side of the evolutionary bargain. Fetal chimps' heads are considerably smaller than mother chimps' pelvises, so there is spare space available when a chimp is born. In this, our closest living relative, the pelvis forms a wide ring through which the baby can pass, and it passes through in the same orientation as most mammalian babies, with its spine facing its mother's spine. Having a brain one-third the size of a human has obvious advantages.

Australopithecus was probably already adapted for bipedal walking, but its brain was not much larger than a chimp's, so there was still a lot of obstetric leeway. The pelvis was now a different shape, however, and we think that baby *Australopitheci* may have been born with their head 'sideways' relative to their mother's body, with their nose pointing towards their mother's left or right hip.

Thereafter, as the brain grew still larger, human birth lurched perilously close to the limit of what is possible. There is now little spare space between the modern human baby's cranium and its mothers birth canal, and that canal has changed from a shallow ring to a kinked tube which guides the fetal head through its tortuous journey. The ninety-degree turn of the *Australopithecus* head has been gradually increased to one hundred and eighty, so that human babies are born 'facing the wrong way', with their bellies facing their mother's spines. In addition, their heads must turn and their necks flex at just the right moments if they are to reach the outside world safely.

The growing human brain made women's lives much more dangerous. In the developed world, women's life expectancy is now longer than men's, but before the advent of modern obstetrics it was shorter. The evolutionary pressures on women's pelvises were immense for most of our species' history – and any genes which caused the 'wrong' pelvic shape were soon eradicated by the cruel hand of natural selection. As a result, human female pelvises are exceptionally large – even before puberty, but much more so after it, when its internal capacity increases at the expense of efficient walking and running. The internal shape and dimensions of the pelvis are among the most rapidly evolving characteristics of the human species, and exhibit striking variations between different human populations. Indeed, the pelvic measurements which show the greatest differences to those of men – the greatest dimorphism – are the ones particularly important for successful birth. The pressure was on, and is always on, to get the female pelvis *right*.

Creating this obstetric masterpiece was a critical phase in the story of the female form, and it effects have spread throughout the body. The connection between the pelvis and the lower spine has been especially strengthened in women, and women's spines even have one more 'wedged' vertebra to allow them to lean further back when they have to counterbalance a swelling pregnant belly. In other words, women's backs are more flexible so they can adopt the 'laid-back' waddling stance of late pregnancy.

Widening the inside of the pelvis necessarily led also to its external widening, and from this time onwards *Homo* women had characteristically wide hips, even before the advent of the curves which were soon to cloak them. Hips have other uses too, and some claim that women's hips also broadened to allow them to perch infants on them. And indeed, humans are unusual in undergoing a phase of infancy when they are too big to snuggle to the breast, yet too little to keep up with their parents on foot. Thus the female pelvis may not be just a bipedal engineering wonder and capacious canal, but a child-seat as well. Throughout human history, all of these pressures have jostled and argued to create today's uniquely, visibly wide female pelvis.

Fat is inseparable from femininity. Females of many species establish patterns of fat distribution which differ from those of males, but this tendency has been taken to new extremes by human females. Indeed, most aspects of the female body which we think of as distinctively womanly – thighs, buttocks, breasts, reduced muscle definition – are the results of female patterns of fat deposition. Fourteen per cent of the average man is fat, whereas 27 per cent of the average women is – and much of that extra fat is located where it can be seen.

Today, most of what we hear about fat is negative. We are constantly told we are living through an obesity epidemic, and that fatness causes disease, yet for the majority of humans' tenancy on Earth, fat has been our saviour. The average person contains enough fat to provide sufficient calories to survive perhaps two months of starvation, and indeed, this is probably precisely what fat was for (for example, it is striking that, following a diagnosis of terminal cancer, fatter people survive longer than thinner people). During our species' precarious past, the ability to store fat was essential for us to cope with alternating periods of plenty and want. Once harvested or killed, food items deteriorate rapidly in the African climate, so instead we learnt to store our calories internally, as non-putrefying, readily accessible fat. Again and again, it was fat which saved us from death.

Small animals are always closer to starvation than large ones. A shrew or mouse's fat store can fuel it for a only few days of its frenetic lifestyle. As animals get larger, however, the economies of scale mean that their fat can provide resources for ever longer periods. Above a certain size they can carry sufficient fat to buffer them against seasonal variations in food availability, and human-sized animals can even store enough fatty calories to sustain them through the most energetically demanding things they ever do: pregnancy and lactation. This is almost certainly one reason why humans became progressively larger throughout their evolution – to shield them from the harsh seasonal fluctuations of their new environment, and especially to allow women to store the resources for bearing and rearing children. Thus humans, both male and female, are big because women had to be big.

Compared to other animals, women store a great deal of fat in anticipation of reproduction. Other female mammals may establish some fat reserves in advance, but many rely instead on their ability to schedule birth at a time when resources are plentiful. Rather than accumulating large stores of internal fat, many animal mothers live off the fat of the land – extracting calories from their environment as they need them, when times are good. This is why herbivore babies are often born in the spring, with weeks of good pasture ahead, and why carnivore babies usually appear just before their mothers' herbivorous prey are at their slowest and weakest. Breeding seasonally is a good solution to the demands of feeding offspring for antelope, voles, wildcats and wolves, but it only works when infants grow up quickly – by the end of the few weeks it takes for summer or the killing season to pass, these species' offspring will already be mature and independent.

In contrast, for human females, reproduction and childcare is a much more long-term investment. Human babies need calories and care for *years* – far longer than one single season of plenty, so women must plan much further into the future, and they do this with their curves. Over the course of puberty, most women lay down between 10 and 20 kilograms of extra adipose tissue, much of it on the buttocks and thighs (breasts are mainly fat too, but we will consider them later). That 'gluteofemoral' fat stays in place, a protected reserve of calories – some might even say 'stubborn' – until it is mobilised during breeding.

Of course, not all women today have babies, but by definition *all* their female ancestors did, and the extreme energetic demands placed on them have profoundly altered the biology of all women. Women differ markedly from men in their patterns of fat metabolism, and they react to hardship in fundamentally different ways. For example, at high latitudes where the air sucks heat from the body and food availability is erratic, human males have adapted by evolving more heat-generating, activity-promoting muscle mass, whereas human females have evolved larger heat-retaining, calorie-hoarding fat stores. Also, there is considerable historical evidence that women are more resistant to famine than men.

Even the specific types of fat stored in female buttocks and thighs may give us insights into our species' curvaceous past. Fat is not a homogenous tissue, and the lipids it stores vary around the body. For example, studies show that women's lower body fat is especially rich in particular lipid molecules (omega-three poly-unsaturated fatty acids), and that this type of lipid is especially important in building and maintaining the enormous human brain. Excluding water, these lipids make up roughly one fifth of the brain's weight, and it seems that the continual torrent of brain-building breast-milk lipids squirted into human babies is largely derived from the fat stored in their mothers' thighs and buttocks. Women with larger thighs have higher circulating levels of these lipids, and there is even evidence that they, and their children, are more intelligent as a result. And intriguingly, the boost to intelligence conferred by these fats appears to be greater in female children than male children.

In this way, women's buttocks and thighs can been seen as living 'fossil' evidence of the importance of storing fats to fuel the growth of children's extravagantly large brains. So now we know that bums make brains, but frustratingly, we still do not know when ancient women actually evolved their distinctively curvaceous fat stores. Fat leaves no fossil evidence – no hint on fossil bones that it ever existed. After death it soon becomes rancid and rots, even if it escapes being picked from human carcasses by calorie-seeking scavengers. However, this has not stopped anthropologists developing mathematical models of human fat evolution based on trends of height, weight and fatness observed in humans and other primates. These models are tentative, and their results must be interpreted with caution, but they seem to demonstrate an unstoppable rise to primacy of female fat. By the era of diminutive *Australopithecus*, our ancestors may have had a body fat content of 10 per cent – already high for a primate. By the time the creatures we now call *Homo* appeared, two or three million years ago, fat deposition had become a predominant feature of our ancestors' biology. By this stage, women probably already possessed much more adipose tissue and much less muscle than men. Curvy women were walking the dusty earth for the first time.

The nuggets of bone meticulously gleaned from the African soil are finally revealing where women's unique bodies came from – why humans reared up to walk on two legs, how the sexes' body shapes diverged so dramatically, and possibly even how women acquired their characteristic hippy, curvy shape. Yet there remains one question which sits uneasily with our contemporary enlightened views of women's status in society: why do women's bodies seem comparatively unathletic?

Relative to men, or other female animals, women appear alarmingly ill-equipped for physical exertion. They possess less muscle and are much less physically strong than men; they have a small chest with smaller lungs and a smaller heart so cannot extract as much oxygen for physical exertion; their pelvis is wide and less mechanically efficient. Many of these sex differences are also present in other species – think of the shapes of male and female horses, or lions – but as we have seen, human females are also encumbered by large amounts of fat. And much of that fat is located in mechanically inefficient places – continually rotating about the body's pivots and levers in thighs, buttocks and breasts instead of being packed into a central, easily portable location. Female athletes struggle to attain the fat and muscle composition of even relatively unfit men. Men outperform women in almost all sports that require strength, speed and even endurance.

It is not so much the idea that women are less physically swift or powerful than men which rankles, but rather the suggestion that this difference may be an indication of women's roles in ancient human societies. And indeed, for a long time anthropologists assumed that women were inherently fatter and less athletic because the demands of pregnancy, lactation and childcare restricted them to an essentially sedentary existence. Men hunted, and women sat around on their lardy bottoms lactating and picking a few berries, or so went male-dominated anthropology's story.

More recently, a newer generation of scientists has challenged this traditional view of women's bodies. They warn that it is dangerous to draw conclusions about social roles from physical sexual dimor-phisms, especially as we do not know which came first – the physical differences between the sexes, or their accepted roles in society. They also allege, probably correctly, that our view of human pre-history has been skewed by the importance we have placed on hunting, and the image of primordial athletic manliness this perpetuates. There are a few human societies in which women hunt – the indigenous Agta hunter-gatherers of the Philippines, for example – but they are the exception rather than the rule, and in primates as a whole it is the males who do most of the hunting. However, it is entirely possible that hunting was not as important a force in human evolution as some testosterone-driven savants might have thought – it may in fact have developed as a rather peripheral, almost recreational activity, perhaps as a by-product of overzealous predator deterrence. The idea that women are adapted for a sedentary lifestyle also presupposes that divisions of labour are consistent across human societies. In fact, the phenomenon of sex-based division of labour *is* omnipresent in our species, but the actual allocation of particular tasks to the two sexes varies dramatically around the world.

Women's roles in ancient human societies may have been neglected by traditional anthropology, but some distinctive features of human female bodies and human female roles still require explanation, even if those explanations do not sit comfortably with our current social philosophies. Women's bodies are less biomechanically efficient than can be explained by the demands of pregnancy or breastfeeding, or the simple fact that women are smaller than men. However, in the hunter-gatherer societies in which our species has spent most of its existence, women indeed were and are unavoidably physically constrained by pregnancy, lactation and childcare. As a result, their contribution to food-acquisition varies – in some societies they acquire no calories, in others they may acquire as many as men, but in all societies their contribution is more erratic. A consistent feature of human life is that men tend to engage in activities which take them further away from the family base and, most importantly of all, human children suffer exceptionally low mortality because their mothers do not usually get involved in more dangerous modes of food acquisition. Unlike many primates, humans rarely forage in all-age, both-sex packs. So maybe women really did evolve, to some extent, to remain sensibly removed from the boyish fray.

The last ten million years have made women's bodies far more unusual than we could ever have expected. Almost every aspect of the female form has undergone profound re-engineering to ensure the survival of our unique, weird species – all of us: men too. And the biological and social roles which came with being a female human have, if anything, conspired to free the female form to be a complex, subtle, powerful, and sometimes paradoxical thing. And that is how the story of curves begins – for our entire species, and, as we will see in the next chapter, for each individual woman.

TWO
Where women's bodies come from

You should not seem to remember that there is such a thing as a man in the world, and you ought to imagine every body to be of the same sex with yourself. I should like to see you give people to understand, that you know that a man has no breasts, and no hips, and no . . .

The Monk, Matthew Gregory Lewis, 1796

'I was rather horrified as my shape changed as a girl. I didn't like having bosoms as I found they were a nuisance. I didn't like becoming womanly, with everyone teasing me. Later I think being attractive to the boys made a difference – that was what eventually made me more confident.'

Anonymous interviewee 'E'

As girls start to become women, they feel their bodies changing around them. Things grow. Shapes alter. Things switch on for the first time. Becoming the only curvaceous animal in existence is not a straightforward process, physically or psychologically, and growing up transforms not only girls' appearance, but also their behaviour, thoughts and feelings. This process of radical feminisation permeates almost every part of the body – far more thoroughly than you might expect.

The millions of years of human evolution which created women's bodies did not leave us with a definitive, finished product. Instead, every time a girl grows up to become a woman, the female form must be reconstructed anew, resurrected. However, the story of human female development can still tell us a surprising amount about why women's bodies came to look like they do. It can also help to explain one of the fundamental mysteries of female bodies: why they *vary* so very much. Women have curves in particular places, and different women curve in different places, and in this chapter I will investigate why this is so.

Humans are mammals, and almost all mammals have one distinctive feature in common: they spend their early existence growing inside a mother. Throughout our most formative weeks we all live in a female environment, awash with female influences, and this is probably why the triggers which determine our sex are usually 'pro-male' – they can actively drag half of us away from this primordial state of female domination.

In humans, sex is determined by the chromosomes – condensed bundles of DNA genetic material – which we inherit from our parents. Of the twenty-three pairs of human chromosomes, one pair is special because its two partners look very similar in women (and are both called 'X', a letter signifying the mystery they presented to early geneticists), whereas the two partners look different in men (a large single X accompanied by a shrunken vestigial chromosome called, with alphabetical logic, 'Y'). Thus women have XX sex chromosomes and men have XY – women get an X from each parent, whereas men get an X from their mother and a Y from their father. Also, we have learnt from people with abnormal sex chromosome complements, such as X, XXY, XXX, XYY, XXYY, that it is the presence or absence of a Y which controls our sex. In other words, if a human has a Y they look male; without a Y, they do not. And it is the Y chromosome which diverts the male-destined half of us from the Y-less maternal environment in which we developed.

In fact, we now know all the biological steps leading from sex chromosomes to actual sex. For example, we have discovered the single gene on the Y responsible for flicking the male switch, and have called it *Sry*, or 'sex-determining region on the Y chromosome'. So once again it is the presence or absence of the male factor *Sry* which dictates our sex. The next step in the sequence is that *Sry* causes embryos to develop testicles instead of ovaries, and those testicles then manufacture the hormones which masculinise genitals, bodies and brains.

It is notable that, throughout this process of development in the womb, all the active triggers are for maleness. If any of these steps fail – the presence of Y, the activity of *Sry*, the formation of testicles, the production of male hormones – then human babies end up looking female. However, despite the concerns of feminist philosophers, the fact that femaleness appears to be the 'default' state and male the 'actively derived' state does not really mean anything as far as the status of women is concerned. We mammals needed a simple binary switch to make our babies male or female, and it seemed sensible to make that a switch-to-maleness, easily detectable inside a gestating female body. Once we escape that body, however, active female forces can come to the fore at last.

And those forces, slowly at first, but with gradually increasing rapidity, make their effects known by causing the deposition of one particular body tissue: fat.

Fat is not the amorphous blobby stuff we usually assume it to be. Body fat does not just pool in amorphous puddles; instead, it is neatly encapsulated in dedicated cells called adipocytes. Adipocytes are unusually large cells, each distended by a droplet of liquid lipid which takes up almost all the space inside the cell. This droplet is surrounded by a thin 'rind' of cellular machinery which controls the entry of fat molecules as and when they become available, and the exit of fat molecules when they are required elsewhere. Fat contains large amounts of usable energy, and it is by far the body's biggest store of calories – humans often carry enough fat to keep them alive for a couple of months, and most women carry more than enough fat to support the demands of an entire pregnancy. Fat is fabulous – it allows us to survive starvation, it keeps heat in, and it even helps us to float.

Fat is hugely important in human life, and it is an active participant in the everyday running of the body. Many hormones cooperate to control the storage and mobilisation of this valuable energy reserve, and one of the most important is insulin, whose main function is to encourage cells to store energy-containing molecules. In the case of fat, this means making adipocytes suck lipids out of the blood and store them in their internal droplets. As a result, adipocytes can swell tenfold, and probably more. In fact, adipose tissue can cope with the most extreme expansion and shrinkage of any tissue in the body.

Intriguingly, fat is also directly wired up to the nervous system. It is so important for fat to be made available when it is needed that this cannot be left solely to the relatively sluggish effects of hormones. Because of this, nerve fibres called, for arcane reasons, 'sympathetic' nerves extend out from the spinal cord and release the chemical noradrenaline directly onto adipocytes. This makes them release fat molecules into the circulation – the opposite effect to insulin. This direct connection between the brain and fat may seem surprising, but as we will see, it is important for the formation of female curves.

As well as storing and releasing energy, it now appears that adipose tissue also sends signals to the rest of the body to announce how much fat remains in the coffers. Adipocytes secrete a hormone called leptin which, among other things, acts on the brain to control appetite. For example, if a mouse is skinny, it will have less fat and make less leptin, and this makes it feel hungry. In contrast, a more plump mouse will have more fat, more leptin, and its appetite will be suppressed. However, the system does not work quite the same in people – low leptin makes people hungry just like mice, but high leptin is not good at making humans feel full, and this may explain why we so easily get fat. However, leptin still exerts diverse effects around the human body, often playing a role in the creation of female body shape.

Even a superficial examination of a woman's body demonstrates that fat is not stored uniformly, but in separate, distinct depots – and we now know that each of those depots has its own particular role to play. This array of different types of fat is actually a characteristic of all mammals, but it has been taken to extremes in human females. For example, samples retrieved during liposuction in women show that different depots contain different sizes of adipocyte (the buttock and thigh 'gluteofemoral' ones are the biggest); store different types of lipid molecules; have different numbers of sympathetic nerves controlling them; and carry different 'receptor' molecules to allow stimulation by hormones. These variations explain why different regions of body fat have different functions – storing lipids from meals, establishing short- and long-term stores, providing fuel during exercise and so on. All fat is not created equal, and therein lies the key to women's distinctive and varied contours.

Compared to other animals, who usually leave the womb in a fairly scrawny state, humans are born unusually fat. There are five particular phases in our lives when we establish fat stores, and the first of those comes immediately before birth. (We will return to some of the other four phases later on: chubby late babyhood, puberty in girls, just after we stop growing taller, and middle age.) In fact, boys' and girls' body shapes already differ by the time they are born. At birth, both sexes possess the same absolute amount of fat, but male babies are slightly longer, and thus heavier, and the extra weight is largely made up of lean muscle mass. As a result, baby girls consist of perhaps 14 per cent fat whereas boys are 12 per cent fat – not a huge difference in baby-bounciness, but real nonetheless. Between the age of six and the start of puberty, boys deposit more muscle and girls deposit more fat, especially on their hips, and girls already have wider pelvises too. And although we do not think of children as sexual creatures, there is evidence that the formation of the female form is coordinated by sex hormones even in childhood – for example, girls with higher circulating levels of sex hormones tend to deposit more fat long in advance of puberty.

During puberty – a protracted process in our species – these sex differences are greatly accentuated, and it is girls who undergo the most dramatic alterations in body shape. By the age of twenty, boys end up with a body composition similar to many other species – with unremarkable proportions of muscle, bone, fat and internal organs – but girls have developed into a curvaceous form not seen elsewhere in nature. Both sexes gain half of their final bodyweight during puberty, and for girls that means sometimes gaining weight at the rate of perhaps 9kg a year. Most remarkably, girls accumulate fat twice as fast as boys, so it constitutes an increasing fraction of their bodyweight as they get older, compared with a decreasing fraction in boys. The end-result of all this is the 10 to 20kg of distinctively female gluteofemoral fat I mentioned in the previous chapter, as well as the fact that the average young man is 14 per cent fat, whereas the average young woman is 27 per cent fat.

Of course, this fat accumulation can change girls' appearance dramatically – softening their contours and hiding muscle definition – but there is a great deal of variation between individuals. The tennis champion Martina Navratilova has recounted how her biology teacher used her well-defined sinews, rather than those of the boys in her class, to demonstrate muscle anatomy, but she is an exception. Some girls gain more fat, some less, but almost all gain *some*. We will return to the question of *why* girls vary so much in their post-puberty shape later on.

Girls acquire their gluteofemoral fat some years before the bony pelvis widens. Thus it could be argued that girls' wide hips give the illusion of having a wide birth canal – they have evolved to look sexually mature long before they are actually fully fertile. This is a striking contrast to boys, who become fertile early, but look immature for an irritatingly (as I recall) long period. There are various theories about why girls evolved to grow up more quickly. Some have suggested that they are laying down energy reserves for later, or that they are delaying fertility until their pelvis is wide enough for a large human baby-head to pass through. Others claim that early maturation allows them to enter the adult social world early, or that their 'apparent eligibility' as a sexual partner once protected them from violence by dominant males. It has even been claimed that early maturation evolved to allow girls to be 'assessed' and 'traded' between ancient human tribes before they had become sexually active.

Whatever the reason, girls' appearance changes profoundly at this time. Often the earliest physical change of puberty is the start of breast development, although this process will take many years to be complete. In all other animals, mammary glands remain small during puberty and do not expand until some way into the first pregnancy. However, in humans the opposite is true, and breast formation *precedes* almost all other aspects of sexual maturation. Also, uniquely, human mammary glands are embedded in two relatively large pads of adipose tissue, giving the teenage human breast a relatively mature appearance. It certainly seems that human female sexual development has evolved to provide obvious visible advertisements of femininity – and that these advertisements are strangely misleading because they develop long before the actual ability to bear children.

The same is true for the buttocks. Fat deposition makes girls' bottoms swell during the teenage years, and the developing curvature of the spine and remodelling of the pelvis enhance the buttocks' visual impact by making them protrude backwards. The human buttocks have a central muscular core, comprised mainly of the large gluteal muscles, but most of their distinctive appearance – their globularity and the distinctive crease separating them from the thighs – is due to a thick layer of subcutaneous fat. Only rarely do humans become so emaciated that the more angular outline of the underlying gluteals becomes evident.

Unlike prenatal female development, in which the absence of male influences passively leads to the formation of the female body, postnatal life appears to be a more active process of woman-creation. And it is in women, not men, that many of the distinctive features of our species – flat face, upright stance, and subcutaneous fat stores allocated to long-term provision of care and calories to offspring – reach their visual zenith. And as we will consider in a later chapter, these obvious, advertised changes can have far-reaching, and sometimes adverse, effects on teenage girls' psychology – as the almost boyish shape of childhood is lost under layers of fat: a substance too often mentioned in the context of obesity, laziness and greed. Yet if we want to understand why women's bodies curve in the places they do, we must understand the hormones which cause fat to accumulate in those curves, and how those hormones suddenly take control at puberty.

Puberty marks the great watershed in the biology of femininity. From conception to adolescence, bodily femaleness resulted mainly from the absence of pro-male influences, but this now changes, largely due to increasing production of the ultimate pro-female hormones: oestrogens. Oestrogens are so important that they deserve a portrait:

This is oestradiol-17β, one of my personal favourites, and although it may look rather jumbled there is a great deal of biology going on in that diagram. Oestrogens are members of an ancient and widespread family of hormones called steroids – insects also have them, for example. They exist throughout biology – and oestrogen-like compounds are even present at high levels in beer, that manly drink. Steroids have a bewilderingly wide range of effects on the body, and oestrogens are no exception. Almost every human cell possesses special receptor molecules which bind to steroids, and by which steroids control many aspects of cellular functioning.

All steroids have their carbon 'C' atoms arranged in three hexagons and one pentagon as in the diagram. Yet oestrogens have additional structural features which make them unusual among steroids. In the diagram you may notice that the lower-left hexagon has some double-links in it, and is not as encumbered by extraneous hydrogen 'H's as the other polygons. This is because, unlike other steroids – progesterone, testosterone, hydrocortisone, for example – oestrogen's leftmost hexagon has been partially denuded of hydrogen, or 'aromatised', and this fundamentally alters its three-dimensional shape. This unusual shape means that oestrogens can exert powerful and specific pro-female effects, even at very low concentrations. In the 1990s I used to study the effects of oestrogens and I often had to measure quantities of oestrogens in blood samples amounting to no more than a few trillionths of a gram.

From puberty onwards, most oestrogens are made by the ovaries, and they drive most of the distinctive changes in girls' bodies. Oestrogens encourage adipocytes to multiply, swell, and accumulate lipids, and are the main reason why girls deposit fat on their upper arms, thighs, buttocks and the 'pubic mound' in front of the pelvis. Good evidence for the effects of oestrogens is that they also cause similar effects in other situations, such as when post-menopausal women use hormone replacement therapy, or when men have diseases, such as hormone-secreting tumours, in which abnormal amounts of oestrogens are produced.

The end-results of oestrogenic stimulation during puberty are clear, but less obvious are the mechanisms by which they induce such a distinctive expansion of particular fat depots. It is likely that they achieve this partly because some depots have more oestrogen-receptor molecules than others. Indeed, the distribution of receptors differs between the sexes, and having lots of oestrogen receptors correlates with hip and bottom size in women. Also, oestrogens may alter the number of sympathetic nerves which plug into any particular fat depot, so that the gluteofemoral depot, for example, is rendered relatively insensitive to the fat-depleting effects of nervous stimulation.

In addition, by the end of puberty, oestrogens have completely realigned female fat metabolism, so that women store and use fats in entirely different ways from men and children. For example, the cocktail of lipid molecules in the blood differs between the sexes; sympathetic nerves eradicate belly fat much more effectively in women; women get more of their energy for exercise from fats than men; and women's leptin levels are much higher than men's for any given level of body fatness.

Oestrogen are even responsible for the alterations to the structure of adipose tissue which underlie that female nemesis: cellulite. In both women and men, subcutaneous fat is internally supported by a mesh of collagen fibres, but in women these fibres are bundled into a more 'coarse' arrangement and are aligned differently – in men, abundant fine fibres run parallel to the skin surface, but in women, sparser, thicker bundles lie perpendicular to it. Because of this, the slightest amount of excess female fat can bulge outwards between the bundles and give a lobulated, dimpled appearance, especially when pressure is applied. And since this is something that occurs below the surface, rather than on the skin, all the expensive 'anti-cellulite' creams in the world cannot restructure this distinctively female adipose architecture.

Wherever one looks, oestrogens have a feminising effect at puberty – they make lips fuller, they dilate blood vessels to make cheeks rosy, and the oestrogen-receptors in the developing female pelvis are crucial in giving the birth canal its unique shape. All in all, oestrogens feminise women far more thoroughly than one might think they absolutely *have* to.

Oestrogens are also responsible for terminating vertical growth. At perhaps eighteen years of age in women and twenty-one in men, oestrogens switch off the growing regions of our bones. As it happens, this effect was originally discovered by studying a man who lacked oestrogen receptors, and who was still growing at the age of twenty-eight.

Other hormones also play a role in female pubertal change. For example, male sex steroids (androgens) such as testosterone are sufficiently abundant in girls to encourage skeletal growth, development of pubic and armpit hair, and maturation of the clitoris, yet girls do not have sufficient androgens to cause a male-like body shape. Some have argued that it is the balance between oestrogens and androgens which is important in body shape, while others argue that the relative roles of the two are not yet clear. Yet oestrogens certainly remain the key feminising influence in women's development, and it is clear that much of this feminising involves placing fat depots at particular locations in the body. It is almost as if the oestrogens are anticipating something.

. . . and that something is reproduction.

For most mammals, including humans, breeding is the most demanding thing they ever do. Producing new little members of one's species takes an enormous investment in energy, time and resources – and this is especially true for humans, whose offspring grow up achingly slowly, and whose brains have an irrepressible hunger for calories. Unlike most organs, children's brains cannot 'power down' when not in use, and cannot shrink when times are hard; they are fussy about which nutrients they use, and burn ridiculous amounts of energy. An adult human brain may use a fifth of all the resting body's energy budget, but a child's may use three-fifths, and a newborn baby's may even top a staggering four-fifths. Once again, the vast developing brain imposes extra demands on human parents which are simply not inflicted by other species' offspring. And we cope with those demands in two ways: fathers are meant to stuff calories into mothers and children, and mothers are meant to have established adipose calorie reserves in advance.

Having said that, pregnancy itself is not as energy-sapping as you might expect. A pregnant woman probably needs an extra 240 calories a day, which is only an extra 15 per cent or so. (I use the word 'calories' in the colloquial sense – what everyone calls 'calories' should, in fact, be called 'kilocalories'. No doubt some marketing whiz thought that a two-hundred-thousand-calorie chocolate bar sounded too frightening.) Most of those extra pregnancy calories are not, in fact, used for the developing baby, its placenta or its accommodation, but instead they are used to deposit yet more fat in the buttocks and thighs. Most pregnant women lay down between 3 and 4kg of additional gluteofemoral fat during pregnancy, although this figure varies a great deal. In fact, laying down fat is not *essential* for successful pregnancy, and an appreciable minority of pregnant women actually lose fat.

Indeed, most women deposit no fat at all in their third trimester, and although it may sound strange, this is the part of pregnancy when women worry least about their size – perhaps the happy reason for their increased rotundity is now so obvious that they are no longer concerned about others misconstruing it. Overall, an average-weight woman should gain between 11 and 16kg by the time she gives birth – of course, much of this weight is baby, fluid, placenta, uterus and mammary tissue, but an appreciable chunk of it is fat. However, lighter women are advised to gain more weight and heavier women to gain less, or none at all. This is because pregnancy can easily conceal insidious, hard-to-shift excess fat accumulation – and in developed countries a fifth of women still retain 5kg of extra fat eighteen months after they gave birth.

Yet all this gluteofemoral fat is not for pregnancy – it is for lactation. Breastfeeding uses far more calories than pregnancy – maybe 750 calories each day, and this level of demand requires a complete reconfiguration of women's metabolism. Transferring nutrients to babies in the form of milk uses much more energy than using a placenta; babies also burn much more energy after they are born – keeping warm, breathing, digesting food, removing waste products. Postnatally, babies also need a lot more parental energy to keep them cuddled and clean. Thus the real demands of human reproduction begin to bite in the days immediately after childbirth, so this is when women finally deploy their not-so-secret weapon. From puberty to birth, the gluteofemoral fat depot has seized lipids from the blood and jealously hoarded them, but now that pattern reverses. Buttock and thigh fat cells now mobilise their reserves and pump lipids into the blood, and most are then absorbed by the mammary glands and converted into milk for a hungry baby.

This is what those distinctive yet previously inactive gluteofemoral fat stores were waiting for – the demands of lactation. Breastfeeding strips calories from the body at an alarming (or wonderful, depending on your point of view) rate, so mothers' body fat content declines and their buttock and thigh measurements decrease. These changes in body shape are only seen in mothers who breastfeed, and they are more pronounced if they give their babies no other sustenance. Mothers' bellies shrink too, but this has less to do with the loss of fat, and more to do with re-tightening abdominal muscles, and the dramatic, sometimes uncomfortable shrinking of the uterus, hastened by the hormones released when babies suckle.

Lactation is so disruptive to the usual pattern of female fat metabolism that it has long-term impacts on body shape. As you might expect, successive lactations permanently increase the pendulousness of the breasts, but they also alter the distribution of lower body fat. The depletion of gluteofemoral fat depots which occurs during breastfeeding is often only partially reversed by the next time a woman becomes pregnant. As a result, women progressively lose buttock and thigh fat with successive pregnancies, and start to store their fat more centrally. Potentially, this depletion means that there is less baby-brain-feeding gluteofemoral fat for later siblings, and this has been claimed to explain why younger siblings tend to have lower IQs.

In our ultra-modern abundance-filled Western world, all this elaborate buttock-to-baby channelling of energy may seem a disconcertingly primal thing, but this is because it has its origins in the long-forgotten 99 per cent of human history when our food supply was unreliable. Everywhere in nature, all the time, young animals are perishing because their mothers cannot feed them enough. Even if they do not die, they can be so irreversibly weakened that they never thrive or produce their own offspring. Human life used to be like that, and women's bottoms and thighs evolved with the express function of preventing these disasters.

Interesting things happen to the shape of the female body later in life. Human development never actually ends, and our genetic developmental clock keeps ticking long into middle age and old age. And women have an exceptional feature built into their genetic programme: menopause. Whereas almost all female animals carry on breeding until they die, women stop producing children during their fifth decade (they would in fact be unique in this respect, had we not discovered that female killer whales do the same thing). As a result, women over forty-five are unlikely to need to breastfeed infants – so they show us what the human female body does when the evolutionary pressure to store fat for procreation is gone.

The average age of menopause is fifty-one or so, but the ability to conceive declines ten years earlier. The ovaries produce smaller amounts of oestrogens, and those oestrogens are often bound into an inactive form in the circulation – although some are still made by other tissues, including fat. The oestrogen decline is probably what causes hips to narrow and waists to widen after the menopause – a reversal of oestrogen-fuelled puberty. Now breastfeeding is a thing of the past, the gluteofemoral stores are depleted and women store energy for their own personal use instead, in similar locations to those primitive beings, men and apes – around the waist in particular. They also store more fat in their upper arms and on their back – thus the elegantly named 'bingo wings', 'love handles' and 'muffin tops'.

If a woman wishes to counter these changes she can lose weight, but this can have unwanted effects. For example, it can exacerbate age-related shrinkage of the fat pads which puff out the cheeks and eye-sockets to give the face its youthful roundness. Indeed, studies show that whereas slight loss of fatness makes people below forty look younger, it actually makes people over forty look *older*. And it is not just changes in fat which cause problems – the skin thins and becomes less flexible, especially under the upper arms and thighs, and gravity continues its unrelenting assault on the weak fibrous capsules of the breasts. Middle-aged humans of both sexes lose muscle mass too, and this can make hands appear bony and faces jowly.

On average, middle-aged women tend to put on weight, mostly in the form of fat, but it is unclear whether this is a result of the menopause, or just a change related to chronological age. After all, middle-aged men put on weight too – and both sexes gain an average of one gram of fat each day during middle age. Weight gain is the body-issue older women complain about most often (as do younger women), but it can, unlike many of the other changes of middle age, be controlled. People who were leaner earlier on in life tend to gain less weight in middle age, or no weight at all, but the opposite is true of people who carried more fat when they were young. However, these are all average trends, just waiting to be bucked by those who, unlike me, have the necessary willpower.

Middle age brings clear biological advantages for women's body shapes, although whether you consider them to be aesthetic advantages depends on your personal point of view. The new, more central location of fat means that locomotion becomes more efficient, without youthful fat depots swinging about with every stride. The tendency to gain weight also reflects a phenomenal new energetic efficiency which saved our middle-aged ancestors in many a famine. It also raises the intriguing possibility that middle-aged women evolved to be *more* physically active than younger women. Today however, most women want to delay or reverse the fat-skin-muscle changes which come with middle age, and I am afraid the best scientific advice is to be virtuous. Be fairly lean when you hit forty, eat sensibly and in moderation, exercise a bit, do not smoke, and slather on cheap moisturiser rather than paying a fortune for creams which come with dubious semi-scientific claims.

From this chapter's brief survey of changes in shape throughout life, it is clear that the female body follows a prearranged path – gaining curves here, losing curves there, as the years creep past. However, although these changes follow a similar general course in all women, one of the most remarkable features of our species is the extent to which its females *differ* in shape.

Contrary to what you might expect, and in comparison to other species, humans do not vary unduly in height or weight. Yet differences in women's curviness can be extreme. Women vary somewhat from the shorter and wider to the taller and skinnier, but the sizes of their various curvaceous bits vary far more, and they also vary *independently* of each other. Some women have small breasts and large thighs, while other women of the same weight can be completely the other way round – there seems to be no obvious link between the sizes of the different curvy bits. In most women one element will be smaller than average, while others are larger than average – and sometimes the differences 'within' one individual woman can be dramatic. Apart from the fact that all women's bodies, even the waifiest, still contain some feminine fat, there seem to be no rules. So where does all this variability come from?

Although variation is an important part of the theory of natural selection, it has proved remarkably difficult to explain why evolution sometimes produces dramatic variability. Because of this we still have no single, simple theory to explain why women's shapes vary so much – and this evolutionary problem will return to haunt us later in this book. One particular problem is that natural selection acts on individuals – deciding their success or failure – but not on entire populations. Thus variability in women's body shapes cannot have evolved because it benefits everybody, but rather because it benefits each individual woman within the population. And it is hard to see how that could happen.

One possibility is that humans evolved so that each woman has the ability to adopt one of a wide variety of body shapes in response to her environment and life experiences – the same way a forest sapling grows into the best shape to exploit the dappled sunlight falling on it. However, although women's life-experiences and diets undoubtedly have some effect on their body shape, it seems unlikely that these influences entirely control the *relative* sizes of breasts, waists, bottoms and so on. Indeed, most women are aware that although they can change their absolute size if they want to, the relative shapes and sizes of their curves remain irritatingly resistant to change – so while this idea might work for saplings, it does not work well for women.

Instead, it is more probable that women's curves are to a great extent predetermined. This implies that the relative sizes of different body regions are genetically controlled, and thus can be inherited. Height, for example, is a strongly inherited trait, and waist, hip and thigh circumference seems to be quite strongly genetically controlled too. Of course this raises the possibility, horrific for many women, that they might inherit their mother's bottom or arms, and indeed this does happen to some extent. However, a woman should always remember that she inherited roughly half of her 'curviness genes' from her father – even though he did not, of course, use them.

Once we accept that body shape is partly genetic and heritable, then there are two specific evolutionary mechanisms which could act to increase the diversity of women's body shapes. The first of these is called 'sexual selection', which I will discuss in some detail in Chapter 4. The second mechanism is 'disruptive selection', in which the environment just so happens to dictate that individuals 'at either end of the spectrum' are most successful. For example, if women with either huge or tiny breasts, but not intermediate-sized breasts, were at some supposed advantage on the African plains, then this could potentially explain why women's breasts are now so variable in size. Disruptive selection is a very real phenomenon, and it may be important in splitting species into two, but it probably does not explain women's varying curviness. After all, there are simply too many intermediately shaped women to fit with the theory that humanity is a species being torn asunder by curviness-based disruptive selection.

For now, probably the most convincing explanation of the variability in women's shapes comes from research which suggests that members of our species are inherently variable in the ways they deal with the demands of life. Studies of populations from the developed world as well as hunter-gatherer communities suggest that, within a given environment, different individuals pursue different reproductive strategies. Some breed early, some late, some produce more children, some produce fewer, and of course all these decisions affect the care lavished on each child. In different communities and different families, men and women vary in their approach to acquiring and storing resources – some are hoarders, some live for the moment, some women gather most of the resources they need, whereas others focus their energy elsewhere while their partner provisions them. And most urgent of all, the spectacular demands of lactation are met in different ways – some women make calories available for breastfeeding by using up their gluteofemoral fat, some reduce their physical activity, others slow their metabolism, and some simply eat more.

In other words, fat is a common currency, and there are many ways of earning, hoarding, conserving and spending it. If people's lives can look different because their financial budgets differ, why might not women's bodies look different because their energy budgets differ?

When it comes to dealing with the challenges that life throws at them, humans adopt an impressively wide range of strategies, and this is especially true of the greatest challenge of all – women's ability to sustain infants. Much of the edifice of the female body exists to bolster this ability. Thus it is because women's bodily strategies for reproductive success vary so much, that their bodies, and the way their bodies change over the course of their lives, vary too.

THREE
The power of curves

The Baron paled at this sight. Candide, seeing his beautiful Cunegonde embrowned with blood-shot eyes, withered neck, wrinkled cheeks, and rough, red arms, recoiled three paces, seized with horror, and then advanced out of good manners. Cunegonde did not know she had grown ugly, for nobody had told her of it.

Candide, Voltaire, 1759

'My least favourite bit is my boobs – without a doubt. They're practically non-existent. I don't feel womanly. I hate them. I despise them. If I could get rid of them and have them replaced, then I would. If I could sell a kidney to pay to get them done, then I would. Taking my top off with my boyfriends has always been a definite no-no. I somehow feel that if I keep my top on they won't notice.'

Anonymous interviewee 'C'

During the last century, female reproduction changed profoundly. In developing countries girls now undergo puberty at a much younger age than they did one hundred years ago. Estimates vary, but over the course of that century, puberty may have been hastened by between three and four years, from an average age of fifteen to eleven. This rate of change is equivalent to roughly *twelve days* of 'hastening' for each year of that century – breakneck speed from the biological point of view, and certainly too fast to be an evolutionary change. What is causing this novel reproductive precociousness?

For some time, scientists have pointed out that this rapid hastening of puberty has occurred during the same period and in the same societies in which children have become better and more consistently nourished, less diseased and less drained by the pernicious effects of parasitism. And because of this, the suspicion has arisen that female fertility is largely controlled by some combination of diet, weight, fatness or body shape.

Most of us have heard of female athletes or women with eating disorders who have stopped having periods, and we often take for granted the idea that women need a certain amount of body fat before they can be fertile. This idea was first proposed in the 1970s and went a long way to explaining the apparent pro-fertility effects of fatness. It provided an explanation for the hastening of fertility in girls, and it could also explain why women with little fat often stop menstruating. It even accorded with the widely held assumption that curvier women – with 'childbearing hips' – were more fertile. In early versions of this theory, it was suggested that girls had to reach a threshold weight of 46kg before puberty could start, although later the threshold became more fat-focused – with a 17 per cent body fat content required for puberty, and 22 per cent needed for regular periods.

The fatness-threshold theory fitted the existing data well. Childhood obesity is indeed linked to precocious puberty and being underweight is linked to delayed puberty, and more moderate variations in pre-puberty body weight also cause similar, but more subtle effects. The theory also seemed to tie in with socio-economic factors relating to diet – a study in South Africa showed that girls raised in relatively privileged families underwent puberty earlier than their less fortunate peers. And indeed, malnutrition, eating disorders and large amounts of physical exercise do delay puberty, and also suppress reproduction in adult women: they stop cycles ('starvation amenorrhoea'), reduce the chances of conception and increase the risk of stillbirth.

Recently, researchers have shifted away from simply measuring body weight, mainly because it is a poor indication of slimness or fatness as it takes no account of height. The most common replacement measurement is the well-known 'body mass index' which is calculated as a ratio:

$$\text{body mass index} = \frac{\text{mass in kilograms}}{(\text{height in metres})^2}$$

Yet even body mass index is not a perfect assessment of fatness or thinness, mainly because it does not distinguish between muscle and fat – very muscly people can be classed as 'obese' according to the ratio. However, it is certainly better than just using weight as a measure, and it will do for now. And indeed, the timing of puberty correlates well with body mass index in girls, although notably this correlation does not hold up well for boys.

The fatness-threshold theory remains appealing, and since it was proposed we have discovered possible mechanisms by which it might work. We no longer think of adipose tissue as an inert, passive blob, but rather as a metabolically active organ which produces and modulates the effects of oestrogen, secretes a wide variety of chemical messengers which influence distant tissues, and also secretes leptin which appears to be precisely what the theory requires – a hormonal barometer of the amount of fat in the body. In addition, the fatness-threshold theory also makes evolutionary sense. We have already seen that human females face extremely high energy demands associated with pregnancy and lactation, so it would be eminently sensible if the biological decision whether to breed was based on the amounts of energy-rich fat stored in the body.

However, this is where problems in the theory start to appear. 'Current' fatness is all very well as an indicator of a girl or a woman's past ability to acquire and store calories, but it does not predict the future. This is especially true of girls at the start of puberty, because so much will have changed for them by the time they become fully fertile – they will have doubled in size and their family and social situation may be completely different. Another problem is that the theory has no 'upper cut-off' to prevent girls and women conceiving if they are obese, despite the fact that being overweight increases the risk of miscarriage, stillbirth and diseases of pregnancy. Finally, it has even been claimed that the theory may have got the whole thing the wrong way round, and that it is early puberty which causes girls to accumulate body fat rather than the converse (although the theory's supporters point out that body mass index at the age of three, or changes between three and six, do correlate well with the timing of puberty).

All in all, the accumulating evidence now fails to support a simple, direct effect of fat on female fertility. Instead, more complex forces seem to be at work. For example, teenage girls who frequently go running are more likely to stop menstruating – yet of those girls, the ones who stop their periods do not have lower levels of body fat than girls who do not stop them. In fact, many biologists now agree that there is so much natural variation in women's weight, fat content and fertility that there cannot be a simple, predictable link between them.

This complexity may reflect the possibility that fertility responds to fatness in different ways at different weights. For example, a direct, causative link between body fat content and reproduction may only take effect at extremely low fatness levels rarely seen in healthy women in developed societies. Thus, according to this argument, a minimum threshold amount of fat may indeed be required to permit fertility, but beyond that minimum requirement, fatness has little effect on most women's month-to-month fertility.

Another possibility is that we have not been sufficiently specific about the type of fat we are measuring. If curvaceous buttock-and-thigh, 'gluteofemoral' adipose tissue is as important for reproduction as I claimed in the last chapter, then perhaps we should be measuring that instead, and ignoring fat elsewhere. And indeed, puberty occurs earlier in girls with more gluteofemoral fat, whereas it may occur *later* in girls with more fat around their waist. Certainly, such discrepancies could explain why whole-body fatness does not reliably predict the timing of puberty. Also, girls who undergo puberty despite being underweight tend to possess more gluteofemoral fat than underweight girls who do not start menstruating.

The gluteofemoral fat depot may be directly measured by various elaborate means, but a convenient estimate of its relative size may be calculated in the form of the 'waist–hip ratio'. The waist–hip ratio is just what it sounds like – the circumference of the waist divided by the circumference of the hips – so hourglass, wasp-waisted and wide-hipped women have low ratios and men with narrow hips and fat bellies have high ratios. And intriguingly, women with low waist–hip ratios have higher levels of oestrogen during their cycles, ovulate more frequently and are more likely to conceive. Thus *distribution* of fat – body shape, not size – is probably more important than total fatness in human female fertility.

As well as the jealously guarded, lactation-earmarked gluteo-femoral fat stores, there is increasing evidence that the female reproductive system also responds to day-to-day fluctuations in energy availability. Energy comes into the body as food, it can be stored in, or liberated from, adipose stores, and it may be expended when women move around, keep warm, pump their heart, or do innumerable other things. This complexity of the female energy budget has led to the idea that fertility is not only dependent on fat stores, but also the interconnecting flows of energy between diet, storage and utilisation. Indeed, it would make sense for fertility to respond to the entire energy budget, rather than just one part of it, because it is probable *future* availability of calories that is important if a child is to be successfully gestated, born and fed. If the energy budget is tight – and it does not matter if that is because of inadequate diet, poor fat stores or excessive expenditure – then the female body interprets this as a sign that it should not breed.

In recent years, physiologists have started to investigate how the energy budget controls female fertility. A region on the underside of the brain, the hypothalamus, and the glandular appendage which dangles from it, the pituitary, have long been known to be master coordinators of reproductive hormones and behaviour, and it now seems likely that they receive many inputs which 'tell' them what is going on in the world of energy metabolism. And this information may come via nerve connections from the rest of the brain, or it may come in the form of hormones.

One such hormone is leptin, which is secreted by adipose tissue. Because it is made by fat cells, it is thought that the hypothalamus and pituitary might use it as an indicator of the body's fat content. With this in mind, it is notable that mice which cannot synthesise leptin are infertile – as if their brains think they are fat-less – but their fertility can be restored by leptin injections. Similarly, puberty may not occur in girls who cannot make leptin normally. Thus leptin probably feeds information about fatness into the brain's decisions about the timing of puberty. Indeed, girls' circulating leptin levels increase before the start of puberty and usually continue to rise until it is complete. And intriguingly, as puberty progresses, buttock and thigh fat produces a disproportionate amount of leptin. The curves are 'speaking' to the brain.

So our understanding of the interactions between body shape and fertility has become much more intricate in recent years – no longer do we think crude fatness dictates whether female repro-duction is switched on or off, but instead the brain uses the subtle, quantitative information it receives to increase or decrease its pressure on the reproductive throttle. In studies in the developing world, reduced diet during the pre-harvest season, or the physical exertion of the harvest itself, has been shown to reduce women's circulating oestrogen and progesterone levels, shorten menstrual periods and lengthen the intervals between ovulations. In developed societies, there is evidence that it is weight *loss*, rather than un-changing slimness, which reduces hormone levels, and it has even been suggested that something as innocuous as regular jogging could do the same.

Once a child is conceived, a woman has little choice but to support and nurture it – effectively, she is biologically committed to it. Because of this, pregnancy and breastfeeding are relatively unaffected by poor diet or low fat stores (except in extreme circumstances). Thus the initial 'decision' to conceive is absolutely crucial, and it is at this all-important decision-point that the reproductive system 'listens' most intently to what the curves have to say.

For most of our evolutionary history, women often did not know where their next meal was coming from, so the biology of the entire body slowly became more and more focused on energy acquisition and storage. Women alive today are, by definition, the descendants of individuals who made the right metabolic decisions during those difficult times, and because of this, a complicated system of whole-body metabolic checks and balances has become etched into their genes.

Today that system still controls women's metabolism, but one important thing has changed. Humans in the developed world now have access to more food than they know what to do with, and this is a situation for which their metabolic control system is not prepared. This is why modern humans so easily become overweight, and we suffer several major diseases as a result. Yet once again, and especially in women, it is often the location of fat rather than its total mass which dictates susceptibility to disease. Having a high body mass index increases a human's chances of dying within the next few years, but having a high waist-hip ratio – deviating from the 'classic' hourglass shape – reduces the chance of survival even more. Thus women's body shape controls not only their fertility, but also their susceptibility to disease and even their longevity.

Of all the diseases caused by being overweight, type 2 diabetes is perhaps the most thoroughly studied (although smoking, lack of exercise and genetic factors are other causes too). It is extremely common in developed countries, affecting people of all ages but especially older people: perhaps one in ten people over seventy years of age has the disease. For reasons we do not entirely understand, in type 2 diabetes, most of the cells in the body stop responding to insulin – the hormone which usually makes them remove fats and sugars from the blood and store them away. The result of this is that blood concentrations of fats and sugars creep upwards, and the pancreas churns out more insulin to exhort body cells to soak up these fats and sugars. However, the body is not responding to insulin any more, so the extra insulin achieves little, and eventually the pancreas simply gives up, exhausted.

Obesity is a major cause of type 2 diabetes, so body mass index can be used to predict the likelihood of someone getting the disease. However, as we have already seen, all fat is not created equal, and the location of excess fat is just as important as its quantity. In women of all ages, abdominal fat – fat held around the waist – seems to be particularly damaging. Both before and after the menopause, large waist measurements and high waist–hip ratios are strongly linked to insulin-insensitivity, abnormal patterns of lipids in the bloodstream and the onset of full-blown diabetes. As a result, pre-menopausal women are relatively protected from type 2 diabetes because they store little fat in their abdomen – completely the opposite of men, in whom the belly is the main store.

The dangers of abdominal fat are only slowly being elucidated, but there is clearly something sinister about this fat depot. Much of the fat in the abdomen is held in a membranous sheet called the 'omentum' which hangs down from the stomach. In abdominal obesity the omentum becomes stuffed with hugely distended adipocyte cells which start to disgorge a variety of damaging substances into the blood – including chemicals more usually seen at sites of inflammation, or free fat molecules overflowing from the swollen fat cells. Because of the anatomical arrangement of the omentum, this toxic mix is delivered by the bloodstream directly to the liver, where it damages liver cells and immune cells. From there, yet more noxious substances now spew into the circulation, from which they enter other body cells and destroy the mechanisms by which insulin exerts its effects.

Thus, by this tortuous route, waist-fatness can destroy the body's delicate energy control mechanisms. Losing weight can help a lot, but some of the damage may already be irreversible. This biological response to obesity may seem entirely counterproductive as far as health is concerned, but it is important to bear in mind that obesity was very rare in humans until the last hundred years or so. Because of this, natural selection has never had a chance to rid us of our self-defeating responses to it.

There is an upside to this sad story of modern life, however, and it particularly benefits women. Just as abdominal fat causes type 2 diabetes, recent research suggests that gluteofemoral fat may actually protect against the disease. Both before and after the menopause, having a 'feminine' low waist–hip ratio, and having high thigh and hip measurements are indeed strongly linked to insulin-responsiveness, normal patterns of lipids in the blood and a low likelihood of getting overt type 2 diabetes. And surprisingly, womanly curves do not prevent diabetes simply by providing a safe alternative to storing fat in the abdomen – instead, the statistics suggest that gluteofemoral fat actively *prevents* the disease in women (but not, perhaps, in men). Not only does thigh fat not release the same toxic cocktail as omental fat, but it may even release substances which promote healthy energy metabolism.

So if a woman is overweight, her health prospects are much better if she retains a pear or hourglass shape. This is no cause for complacency, however, as age brings with it a tendency to shift that previously beneficial fat to a new dangerous, abdominal location.

Having a dysfunctional insulin system can have serious knock-on effects and can cause several other important diseases, especially cardiovascular disease – which accounts for roughly one-third of all deaths in developed countries. Cardiovascular disease comes in two overlapping forms – damage to the heart and brain due to blockage of the coronary or cerebral blood vessels, and high blood pressure – and both of these conditions are strongly related to body shape. There is a clear link between body mass index and cardiovascular disease, but having a non-curvy waist–hip ratio predicts susceptibility to the disease more accurately, suggesting that abdominal fat is once again the villain of the piece.

Type 2 diabetes increases the chances of blockages forming in the coronary vessels because it inflames their inner vessel lining, and it also fills the blood with the fatty building blocks of the clotted plaques of atherosclerosis. In addition, diabetes can damage the nerves which regulate the activity of the heart and the springiness of blood vessels. Body mass index and especially waist–hip ratio are also clearly linked to long-term kidney disease, which can further damage the body's ability to regulate blood pressure.

Yet again, the feminine pattern of fat distribution comes to the rescue. There is evidence that gluteofemoral fat can protect against cardiovascular disease, and it is associated with a healthy balance of lipids in the blood. Indeed, there seems to be no limit to its protective effects – studies show that having a 'womanly' low waist–hip ratio can also protect against gall bladder disease, some forms of breast cancer, and dementia.

The story is not so straightforward when it comes to skeletal disease. It may come as no surprise that arthritis is strongly linked to obesity, because being heavy obviously puts more strain on the joints. However, fat also releases chemicals and hormones which probably directly accelerate the degeneration of joints. At present, there is no clear evidence that distribution of fat in different locations makes women either more or less likely to suffer from arthritis, but I expect that this evidence will appear within the next few years.

However, there are other skeletal problems of particular relevance to women, and which are affected by body shape. One of these is osteoporosis, the loss of bone mineral which occurs later in life in both sexes but is more likely to lead to fractures in women because their bones are less robust to start with. Osteoporosis rather bucks the trend of the rest of this chapter, as studies suggest that having a high waist–hip ratio correlates with good bone density – so having a male-like silhouette may actually protect against this disease.

Another set of female orthopaedic problems clearly related to body shape occurs in women with large breasts, or gynaecomastia. The weight of large breasts can completely unbalance the musculoskeletal system, and lead to upper back pain, lower back pain, shoulder pain, neck pain, arm pain, bra-strap abrasions, and may even compromise women's ability to breathe. Considering its adverse effects, gynae-comastia is surprisingly common, and it is hard to imagine why natural selection has not expunged it from the human population. However, modern surgical breast-reduction techniques offer an almost complete cure.

Many women with larger breasts also worry that they are more likely to develop breast cancer, and at first sight this fear seems reasonable. After all, the last few decades of oncology research have shown that tumours develop from single, aberrant cells which slip free from the restraints which usually prevent their uncontrolled proliferation. So because some women have many times more breast tissue than others, it might seem sensible to assume that those women are many times more likely to develop breast cancer – simply because they possess more individual cells which might grow into tumours. Indeed, this is often assumed to be the case, even by evolutionary biologists, who have sometimes proposed it to be one of the selection pressures which prevented women's breasts evolving to be even larger.

It should be easy to show a link between breast size and cancer risk – woman's breast sizes vary dramatically, and breast cancer is a common disease, and variation and commonness usually make demonstrating something statistically straightforward. In fact it has proved surprisingly difficult to demonstrate such a link. Initial studies suggested that such a simple correlation did indeed exist, but this may have been due to the confusing influence of body weight. Being heavier is already known to be linked to an increased incidence of mammary cancer, but heavier women also tend to have larger breasts – so simple correlations do not mean that breast size *in itself* increases the risk of cancer. However, in women in general, breast tumours are more common in the left breast, which is on average larger than the right. More recent studies have provided evidence of a link between breast size and mammary cancer, but it seems surprisingly subtle – one study suggested that women with larger breasts at the age of twenty were more likely to develop breast cancer later in life, but the correlation only held true for women with lower body mass indices. Another complicating factor in all this is breastfeeding, which changes the shape and size of the breast, but is also suspected of protecting against the development of mammary cancer. However, it now appears that any protective effect of breastfeeding may be lost after the menopause, so it may only actually reduce a woman's chances of getting mammary cancer early in adulthood when she is relatively unlikely to get the disease anyway.

Intriguingly, genetic research carried out in the last few years has now suggested mechanisms by which breast size and breast cancer may be linked. Breast size is a trait which can be inherited, and breast cancer risk is often heritable too – but only recently has evidence appeared hinting that the two may be inherited together. First, scientists discovered seven genetic markers linked to breast size, but then found that three of them are inherited along with, or are physically adjacent on the chromosome to, genetic markers of breast cancer. Three out of seven is too many to be a coincidence, so it now seems very likely that mammary size and mammary cancer are indeed genetically linked in some way. These results do not tell us what these genes actually do, nor unequivocally prove that breast size increases tumour risk in the real human population, but they are certainly intriguing.

Yet still, all of this may seem strangely vague. Some women's breasts are at least ten times larger than others', so why, if a link between size and cancer risk exists, has it been so difficult to find? One possibility is that most of a mammary gland is fat, while relatively little consists of the gland cells which can grow into tumours – so breast size may be a poor measure of the pool of potentially tumour-forming cells. It is also likely that breasts and breastfeeding are so important for human survival that a complex system of checks and balances exists to control cell growth, proliferation and activity in mammary cells. After all, breast cells are unusual because they are *meant* to undergo periodic episodes of frantic cell replication – in preparation for lactation – and I suspect it will be some time before we fully understand how this natural, healthy proliferation is normally prevented from escalating into something more malign. Evolution can produce amazing things, but those things can be fiendishly difficult to understand.

Our journey through the biology of female shape has shown that the elements which make the female body look so distinctive are not just superficial adornments, added to the human form as an irrelevant afterthought. Instead femininity is spread throughout the body, sometimes in places where you might not expect to find it, such as the morphology of fat cells or the angle of the elbow. And it has had, and continues to have, profound effects on human evolution, development, fertility, health and disease.

Many women feel that being a larger or curvier shape is an essential part of their self-confidence, self-belief and self-determination, whereas others feel guilty about being that same shape because they live in a society which often seems to equate fat not with a state of healthy nourishment, but with disease. Of course, the longing to be free to inhabit a body of any shape is usually tempered by a wish to be healthy, but only recently has it become clear just how important it is to treat shape and size differently. After all, we have now seen that rounded hips, buttocks and breasts could even be beneficial for women's health. Millions of years of evolution have ensured one thing – it's not the size of the curves which is important: it's where they are.

Although my intention in the first third of this book was to stick to the basic biology of women's body shapes, that biology has already spilled over into psychology and emotion – confidence, guilt, happiness and desire. In the second part of this book, I will surrender to the inevitable lure of the mind and broaden our story to consider how it interacts with the female body. I will show how all the biology is only our first step on the way to understanding our strange relationship with women's bodies – and as we will now see, one of the strangest stories in human evolution has been the remarkable ascent to primacy of a single organ: the brain.

The girl stood up and stretched her long straight limbs. She felt the distant heat of the ochre sun warm her dark skin and smelled smoke from the previous night's fire in her nostrils. She cupped her breasts in her hands. They seemed to be getting slowly larger ever since the wriggling thing in her belly had appeared. She could not explain why, but this made her laugh out loud.

PART II

THE MIND

'It would be very different being in a man's body – I think it would make me act as a man would act. I'd be more "rigid" if I was a man. Somehow I guess I've learnt from other women how to sit and walk.'

Anonymous interviewee 'A' (age 21, body mass index 22.7)

'When you're a woman and you walk into a professional, serious, corporate environment, you know that every woman is working out whether you're a threat and every man has made an instant decision on whether he wants to sleep with you. You never get the anonymity that men get – I envy them that. You're always making a statement with your body.'

Anonymous interviewee 'B' (age 32, body mass index 26.1)

'I exercise every day. I run 5k at lunch hours, I go to the gym at least five times a week – spinning, fitness classes – I walk everywhere. I'm never going to be the tallest, curviest or prettiest, so I've always wanted to be the smallest person in the room.'

Anonymous interviewee 'C' (age 33, body mass index 21.3)

'I feel guilty about most things I eat. I remember it used to be that everything I ate I thought, "You big fat cow – you've got no willpower." I would feel terrible about myself because in my mind I felt I should have been strong enough not to eat. I used to weigh myself every day – or more often. It would rule my life. But do I ever look in the mirror with no clothes on? God, no. Well – I have done it in the past just to shame myself into going to the gym.'

Anonymous interviewee 'D' (age 40, body mass index 24.5)

'When I look in a mirror I put on my "mirror face" and see someone I know and someone I quite like. If I get "caught" in a mirror, it's a different reaction – I feel I have a downward face, I think I look like my mother.'

Anonymous interviewee 'E' (age 70, body mass index 24.8)

FOUR

What men want and why it doesn't matter

The mode of passion is born of unlimited desires and longings . . . and because of this one is bound to material fruitive activities.

Baghavad-Gita

'I'd much rather have bigger breasts than smaller breasts. I like the power of the female form over men, and using it to get what I want.'

Anonymous interviewee 'A'

A few minutes spent at a newsstand provides evidence enough that women and men have different ideas of what makes female bodies beautiful. Women's magazines usually have women on their covers; men's magazines often do too, but those women do not look the same. Their bodies differ, often their faces and expressions differ, and their poses differ.

For now I will put aside the thorny question of why heterosexual women might want pictures of other women on the covers of their reading material – we will come back to this later – but it seems obvious why heterosexual *men* do. Yet what still requires explanation is why, exactly, men are so visually fixated on the female form, and how they acquire their particular visual preferences. Men like looking at beautiful female bodies, and studies show that men are especially keen to date women whose bodies have already been endorsed as attractive by other men. In more scientific terms: visual assessment is a major determinant of the initial stages of mate choice by male humans.

So why is the *look* of the female body so important? The male quest for female beauty is so deep-seated that it must indicate something more than just the pathological superficiality so often ascribed to men. And indeed, everywhere else in the animal kingdom we find individual creatures behaving the same way – lured by the shallow visual advertisements of the opposite sex. Thus human men are certainly not unique in seeking visual evidence that potential lovers may be 'the one', or at least 'a one' – a future partner with whom they can most confidently pass on their genes.

Yet surely beauty is, at best, only an indirect indicator of how good a co-parent someone will make? Indeed, in the modern developed world the evidence is patchy. Some studies suggest that beautiful people are no more likely to have children or be healthier than less beautiful people, whereas others claim that individual components of beauty are very strongly linked to genetic health and vigour. In particular, there is evidence that more attractive people are cleverer – up to twelve IQ points cleverer in some studies – than less attractive people. Maybe the man at the bar who looks you up and down really *is* only interested in your mind.

In this chapter I will examine what attracts heterosexual men to women's bodies, and why. I appreciate that, after three chapters looking at the grand forces which have forged the female body we see today, you may consider this a rather tawdry way to start investigating the immense power which the female body exerts over the humming loom of the human mind. However, male desire is important because it has been a potent force in the evolution of women's bodies, and today that desire still holds almost half of the human population in its thrall.

Some of us may not look like it, but each human alive today is the product of thousands of generations of heterosexual lust. We are here because again and again, nameless men and women liked the look of each other, and liked it enough to want to have sex. Simply surviving and being successful was not, in itself, sufficient. To contribute to the gene pool of future generations, each of our ancestors also had to find someone to make babies with. Being alive was not enough – ancient humans had to be *wanted* too.

Charles Darwin realised that this was a big problem with his theory of evolution by natural selection. Everywhere in nature there are animals which have survived and prospered, yet are never selected as anyone's mate – many deer stags, for example, make it to a healthy adulthood, yet never sire a calf. Indeed, it is now clear that natural selection has a racier counterpart to account for this – sexual selection. Sexual selection means that animals produce lots of successful offspring not only because they are able to survive, but also because they possess traits which make them more likely to acquire sexual partners. These traits might allow them to compete with members of their own sex – as in the evolution of the human fist I mentioned in Chapter 1 – or they might make them more likely to be selected as mates because they are, well, more *attractive*.

One important implication of sexual selection is that species may acquire many of their distinguishing features because they are attractive to the opposite sex, rather than because they are of any practical use. For a characteristic to be sexually selected it must presumably exist in a rudimentary form in the first place, and then be interpreted as a sign of vigour by the opposite sex. The peacock's tail is biologists' favourite example of this, but what if the same were true, for example, of women's buttocks? If sexually selected traits *are* practically useful too, then all is well, but even if they are not useful they can still win out – at least until the moderating hand of natural selection eventually intervenes and eradicates the peacocks with the most ridiculously extravagant tails, or the women with the most cumbersome bottoms.

Of course, sexual desire takes two, and this is why sexual selection can be such a powerful thing. If, for example, feminine buttocks really are an indicator of health and fitness, then men who like them will end up siring offspring with a better chance of survival. Thus the genes for 'buttock-liking' (which unfortunately have not yet been identified) would themselves confer a beneficial trait which is propagated in future generations, along with the 'counterpart' genes for curvy buttocks. Rather pleasingly, what results is not a genetic arms race, but a genetic love-in – as genes which cause attractive characteristics, and genes which instil a desire for those characteristics, spread and become more potent. Women's buttocks swell; men like them; everyone is happy.

This may sound simplistic, and not very politically correct, but the scientific evidence for sexual selection is extremely strong. Despite the fact that it can be difficult to tell which characteristics evolved by natural selection and which by sexual selection, it is very likely that elements of human female body shape evolved as a result of the latter. This is why human males' desire is so important if we want to truly understand human females' shape.

Before we look at what specific things men find attractive about women, there are two aspects of sexual selection which deserve mention. The first of these is that the sex which contributes most to childcare gets to sexually select the other sex. Peahens lay eggs and look after peachicks, while peacocks prance around being generally unhelpful. This may seem unfair, but it does mean that peahens hold all the cards, as it were – they put in most of the effort, so they can be as dowdy as they like yet still get to select the sexiest peacock. By contrast, everything is much more egalitarian in humans. As we have seen previously, humans are unusual among mammals because paternal investment is extremely important for the success of our offspring, and because of this human males are unusually choosy about their mates. Most male animals will mate with anything that looks vaguely female because copulation will be the end of their contribution to the next generation, whereas human males have to be careful about the individual with whom they are going to co-parent for the next few years, or even decades. As a result, unlike most female mammals, female humans are subject to powerful sexual selection – just like men. They get all that help with the kids, *and* they get desired too.

A second feature of sexual selection is also especially relevant to the human female body. You might expect that a few thousand years of sexual selection by men would result in every woman having a body like Beyoncé or Bardot – yet this does not seem to have happened. In fact, far from making the 'desired' sex all tend towards one single ideal, sexually selected traits often exhibit much more variation than naturally selected ones. There is probably only one good way to make a woman's spleen, and this is why they all look rather similar – whereas if breasts, buttocks and waists are indeed sexually selected traits, then this could explain why those things vary so much. In previous chapters, I did not come up with many convincing explanations for the extreme variability we see every day in women's body shapes, but sexual selection might now explain that variation for us. Perhaps men have spent the last few hundred thousand years sexually selecting mates for all sorts of different reasons, or maybe they are just inherently catholic in their tastes.

Hopefully a detailed examination of male sexual desire will help us discover whether it really is men who have made women such a beautiful and diverse bunch.

Reproductive theorists enjoy making generalisations. They believe that all animals – including humans – have similar aims when choosing mates. The drive to propagate one's genes in successful offspring constrains us all into a remarkably restricted array of likes and dislikes when it comes to the opposite sex. These biologically determined desires make us uncomfortable when they are applied to humans, but they persist nonetheless. For obvious reasons, in this chapter I will focus on the things that heterosexual men look for in women (and especially their bodies), but many of these general principles apply to heterosexual women's desires too.

The 'big three' desirable features for a heterosexual man's mate are perhaps unsurprising – femininity, health, fertility. Men seek women who look obviously like women; men seek women who have good genes and are healthy; men seek women who look like they have years of efficient breeding ahead of them. Of course, these three desires may overlap in complex ways, but they all make sense to a male keen to pour lots of his genes into the next generation.

The 'additional two' lusted-after traits – genetic difference and similarity – are more confusing, because they are opposites. Of the two, genetic difference is the most straightforward: it seems sensible for men to want to breed with women who are genetically dissimilar to themselves, to avoid inbreeding. This is a pressing problem because we all carry damaged genes – every one of us may inherit from each of our parents roughly ten mutations which could potentially cause lethal genetic diseases – but luckily most of us also inherit a 'good' copy of each of those genes from the other parent. However, if men mate with closely related women, their children are far more likely to end up with *two* copies of one of those damaged genes, often with disastrous consequences. Desiring genetic difference safeguards against the calamity of inbreeding.

Despite this, seeking mates who *share* characteristics, and thus presumably genes, might also promote the future survival of men's own genes, especially if those genes are particularly good at promoting survival in the local environment. Indeed, there is good evidence that heterosexual humans achieve this genetic-similarity-seeking by 'imprinting' on their 'other-sex' parent as a model of desirability. For example, studies show that adopted children end up preferring mates who look like their adoptive other-sex parent rather than their biological other-sex parent. This may make for disquieting reading – no one likes to see the words 'parents' and 'mates' in the same sentence – but the fact remains that humans seek a delicate balance between genetic similarity and difference in their sexual partners.

Intelligence

But enough of generalisations about male desire. What *exactly* do men look for in a woman?

For the purposes of answering this question, it is fortunate that the inherent sexism of science means that there has been more research into what men find attractive about women than vice versa. And these studies show that there is a central biological core of non-negotiable heterosexual male lust. This core can be modified and supplemented by cultural influences, but it remains stubbornly present. For the rest of this chapter, this is all going to be very universal and very primal.

In fact, there is only one thing that a man looks for in a woman that is *not* related to body shape – intelligence. Many researchers claim that this has been scientifically neglected at the expense of studying more 'superficial' characteristics, but I suspect this is simply because men's reactions to intelligence and charm are harder to measure than attraction to physical attributes. Anyway, it makes sense for men to find intelligence attractive, as intelligent women are likely to make more sensible decisions about caring for their children in the future. Intelligence is also a characteristic which implies that a woman has herself received good care and abundant resources from her parents, and has thus hopefully inherited good parenting skills. In addition, studies show that intelligence correlates well with general health. Indeed, some have claimed that art, music and humour evolved expressly so that humans could demonstrate their intelligence, and therefore their implied vivacity, to the opposite sex – which may explain why pop stars receive so much sexual attention.

In addition, men certainly find some behaviours attractive in potential mates, but these are even harder to quantify than intelligence. Various studies suggest that men actively seek signs of emotional stability, nurturance, dependability, sexual fidelity, and probably some similarity with themselves in personality traits. They also look for evidence that women are interested in *them* – playfulness, for example, seems to be identified as a sign of both youth and sexual engagement. Women's smiles activate men's orbitofrontal cortex, in the same way that many aspects of visual attractiveness do, so men probably do think women are prettier when they smile. And of course, women's voices are attractive to men too – their higher pitch is a badge of femininity indicative of a small larynx, resulting from low androgen concentrations at puberty. Studies even show that men find women's voices more attractive around the time of ovulation, when oestrogen levels and fertility are at their peak.

Aroma

The second set of features which men find attractive is, perhaps strangely, related to smell. There is a great deal of research which shows that many animals – rodents, fish and birds, for example – select their mates largely on the basis of their odour. In these species, individuals secrete odorants (smelly chemicals) which vary according to the presence of certain genes important in the immune system. A female mouse, for instance, deliberately selects mates that smell like they have different immune genes to her, so that her offspring will inherit a varied mix of genes from their two parents. In other words, seeking a genetically dissimilar mate can make your children more able to fight disease. And remarkably, experiments with pre-worn T-shirts suggest that women do precisely the same thing, and seek out immune-different men. Research even suggests that women married to immune-similar men are more likely to be dissatisfied with their relationship and have affairs.

The evidence that men use smell to choose their mates is less clear-cut, probably because these visually obsessed creatures tend to *see* women before they *smell* them. However, odours can affect many different areas of the human brain, and it is likely that men subconsciously find women's smells attractive in a variety of ways. A small amount of evidence suggests that they too may select partners on the basis of immune-difference, and that they are well able to discern genetic similarity and difference in general. For example, human subjects can often identify their other-sex siblings by smell, and they can sometimes identify their same-sex siblings too; they are less able to detect the smell of half-siblings, and even worse at detecting their step-siblings.

There are also intriguing indications that men can smell the hormonal basis of femininity itself. Oestrogen-like odorants activate men's hypothalamus, an area on the underside of the brain involved in sexual behaviour, and smelling these odorants even makes men think women's faces *look* more feminine. Indeed, smell may be strongly linked to vision in complex ways – for example, women with symmetrical faces are more likely to be rated as pleasant-smelling by panels of blindfolded men, especially around the time of ovulation.

Youth

The third element of female attractiveness is the one that causes the most argument: youth. Evolutionary theory dictates that men should seek women with the greatest *future* reproductive potential – because a man who hooks up with a young woman can exploit many years of future baby production. These women should also be obviously fertile, however, and this is why studies show that most men find women in their early twenties more attractive than teenagers. Yet these are not simple urges, as analyses of lonely hearts ads show that men in their late teens prefer women slightly *older* than themselves, and that this preference wanes until, at roughly twenty-three, they desire women who are similar in age. From then on, the preferred age gap steadily increases in the other direction until men in their seventies seek female partners ten or fifteen years younger than themselves.

This much maligned male quest for youth actually has rather comforting origins. In contrast to most other male mammals, human males seek long-term pair-bonds with females, probably because human children need bi-parental support for extended periods of time. Thus it makes sense for a man to find a woman who is at the start of her child-supporting phase, because he intends to stick around. By contrast, chimps breed promiscuously and do not pair-bond, so males show a preference for older females because they are more experienced at caring for the babies with which these simian lotharios impregnate them.

Some may find it hard to accept that men's attraction to younger women is a reflection of their compulsion to relationship fidelity, but the evidence is all around us. In all human cultures, heterosexual men are, on average, older than their partners. Indeed, male homosexuals seek younger partners too, so this seems to be a general feature of the male brain. A Swedish study showed that women whose partners are four years older than them produce more children than women whose partners are closer to them in age, and a US study showed that teenage girls are less likely to use contraception or have abortions if their partner is more than six years older than them. Clearly, these preferences are built into all of us.

Some of the cues men use to detect women's age may be behavioural or intellectual (or they may simply ask them), but visual attributes are probably more important. Research shows that skin smoothness is a strong cue of youth, as are thick, non-greying hair and healthy teeth. And as far as the body is concerned, men are adept at noticing the age-dependent changes in breast shape, waist size and the evenness of skin tone and texture I discussed in Chapter 2 – however, in all these respects, different women may appear to 'age' at different rates. Yet for the average instinct-driven man this variability simply may not matter. After all, reproductive theory suggests he may not be as interested in absolute chronological age as he is in future reproductive vigour. We might also expect human males to naturally adapt their age preferences to their partner's advancing years, to encourage them to maintain the pair-bond and complete their investment in their children.

Face

With the fourth set of characteristics men use to determine attractiveness, we come to something which overlaps even more with body shape – facial beauty. Contrary to what women might think, men do spend a lot of time looking at their faces, and recent advances in computing have made this a rich area of research. Artificially constructed facial images can now be tweaked and distorted prior to being shown to men and asking how attractive they find them – and this has at last allowed us to dissect the elements of female facial beauty.

For example, facial symmetry is, in itself, attractive. While slight asymmetries seem to be acceptable, the evidence is clear that faces computer-morphed for symmetry are considered more beautiful. If a woman's face and body are extremely symmetrical, then this suggests that flawless genetic and developmental processes moulded her, so it is only natural for men to want to get some of that flawlessness incorporated into their future children.

In fact, the 'genetic stability' implied by symmetry goes further than that, as studies indicate that symmetrical people are healthier throughout life, are more intelligent, suffer fewer psychiatric disorders and even have more attractive voices. The human brain is extremely good at detecting asymmetry, especially in faces. But symmetry comes with a catch: it can make faces seem unemotional or even boring. This may be why women often choose to have extremely asymmetrical hairstyles, or even beauty spots, to contrast their attractive symmetry with a lopsided artifice of quirkiness and vivacity. I also suspect that eye symmetry is especially important for beauty – after all, we all spend a lot of time gazing into each other's eyes, and constructing something as wonderful as a pair of human eyes requires a bewilderingly choreo-graphed intermeshing of developmental processes. So when a man says a woman has beautiful eyes, he is probably *not* saying it just because he cannot think of anything else to say.

It has been suspected for some time that 'averageness' is another component of female facial beauty – having all facial proportions and features of a moderate size, with none exceptionally large or small. While one can easily think of beautiful women with larger noses or smaller chins, for example, computer modelling studies suggest that the desire for averageness is a very real phenomenon. When presented with images of real female faces as well as composite faces computer-blended from several images of those same women, men usually prefer the composites, the 'averaged-out' faces. This may seem like a tendency to prefer blandness, but some researchers believe that it is as important, if not more important, than symmetry. Of course, in real life the two things overlap – an average face is, by its nature, fairly symmetrical, and symmetrical faces represent the average between wonky extremes.

A lesser contributor to facial beauty is that men prefer women whose features share some similarity with their own – and I mentioned previously that this may be caused by 'imprinting' on their own mothers during their boyhood. We may worry about men who date women who look like themselves, and even more about those who date women who look like their mothers, but once again computer studies confirm that this disquieting desire for similarity does indeed exist. If elements of a man's own face are blended into an artificially constructed female visage, then he will tend to find it more attractive. This may be a weak effect, but nonetheless it is an example of men seeking similarity – in contrast to the odour data which suggested that they seek difference.

Facial femininity is an extremely important contributor to men's assessments of beauty: they want women to *look* like women. In previous chapters I described how the dramatic physical differences between women and men evolved – how sexually dimorphic our species has become. And today, men's desire for overtly feminine women may result from aeons of sexual selection making women look womanly and, as a consequence, men have accumulated genes which make them desire that womanliness. However, being attracted to obvious femaleness may also have other advantages – it allows men to direct their sexual interest towards mature women, and especially women who are vigorous enough to maintain the luxury of these 'non-essential' traits. In addition, there is evidence that women with exaggerated female traits develop sexually at a younger age, so that men who select very feminine women may be able to benefit from their unusually extended child-producing life.

There are many characteristics which men's brains identify as feminine, but they fall into three main groups. The first group includes features which women retain from childhood but men do not – such as having a small nose, a small chin, large eyes, thin eyebrows and prominent lips. The second group consists of characteristics which are driven by female sex hormones – including rosy lips and cheeks (although these may also indicate sexual arousal). The third group

includes features which result from a lack of male sex hormones –
an oval, upward facing, 'open' facial morphology with large forehead
and small jaw, and also pale skin. Androgens tend to make facial skin
darker, so men prefer women with pale facial skin which contrasts
strongly with their features.

Apart from symmetry, averageness, similarity and femininity,
we are left with one more, rather rag-bag set of facial characteristics
which men find attractive, and each of which, we suspect, sends men
a specific signal. For example, men find women with dilated pupils
more beautiful, probably because it suggests sexual arousal, and this
is why women, from the time of the Babylonian Empire to mid-
nineteenth-century Paris, used to artificially dilate their pupils with
the poison atropine (belladonna – Italian for 'beautiful lady'). Also,
unlike skin smoothness, which men use as an indicator of youth,
studies show that even skin *tone* – something explicitly promised by
the manufacturers of cosmetic foundations – is used by men as an
indicator of not just youth, but health too. In past times when viral and
bacterial skin diseases were commonplace, this may have been of even
greater importance. It is no coincidence that in medieval England,
milkmaids, who tended to have acquired immunity to smallpox due
to transient minor infections with cowpox, were often called 'pretty
maids'.

Lower curves

By the time we reach the fifth and most important component of male
desire – lower body shape – it is already clear that heterosexual men
come pre-installed with a powerful set of drives to help them select
healthy, fertile, feminine mates. However, it is in their inquisition of
the female body that men demonstrate their greatest deductive powers.
In recent years there has been a flurry of research into this area, to
the point where individual genes related to assessing attractiveness
have been identified, as have specific regions of the male brain which
respond to, for example, images of nude female bodies.

Some researchers claim that the most important criterion of

female body shape is body mass index, which is a reasonably good indicator of fatness. We think fatness is important to men for several interlinked reasons. First, subcutaneous fat is a sign of femininity because, as we have seen, human females deposit large adipose stores to meet the demands of gestating and rearing children – especially their children's large and lipid-rich brains. This is why human female fertility turned out to be so dependent on the flows of fatty calories through the body. So a man who seeks a fat-laden lady may simply be seeking a feminine, fertile partner who is well provisioned for future baby-making. However, being well provisioned says even more about a woman than that – it also implies that she is intellectually or metabolically able to acquire and store valuable calories, or that her parents did a good job of acquiring those calories for her. Either way, she is evidently full of energy – literally – and probably full of good genes and sense too. And indeed, some brain scan studies suggest that body mass index is indeed the best way to stimulate the appropriate parts of the male brain.

There are, however, problems with the idea that body mass index is the main thing men look for. First of all, how is your average chap meant to calculate this esoteric ratio? It may be clinically useful, but I doubt that the male human brain can measure it directly. Some have suggested instead that men are more swayed by women's percentage of body fat, but that seems like an even more difficult thing to measure at a glance. Moreover, men do not simply like women with high body mass indices, but rather they prefer women with moderate indices – in the same way they like 'average' faces. At present, Western men prefer women with near-average or just-below-average body mass indices, but that tells us little because averages change over time. As we will see in a later chapter, women's average body mass indices have changed over the course of history and also vary dramatically around the world today – and men's preferences for fatness have been similarly inconsistent.

A second measure of body shape seems more convincing as a trigger of male desire, although it too is imperfect. This is the waist–hip ratio I mentioned in the last chapter – the circumference of the waist divided by the circumference of the hips. I realise that ancient male hunter-gatherers did not possess tape measures, but the ratio can be estimated by dividing the waist *width* by the hip *width* too, which is something I can imagine the male human brain could easily achieve. Wide hips and narrow waists are easy to *see*. This measure of curvaceousness has the advantage of indicating shape, not absolute size; it is independent of height, and as a result it may reflect a woman's underlying metabolism and hormones rather than how much food she happens to have eaten recently.

Until puberty, the waist–hip ratios of girls and boys are similar, but once sex hormones take effect women's hips outgrow their waists dramatically so that a woman's ratio is usually less than 0.80 whereas a man's is usually more than 0.85. Many teenagers wear low-cut trousers nowadays, but only teenage boys like them to look like they are falling down, demonstrating how masculine and hip-less they are. Conversely, men like women to have low waist–hip ratios because they indicate a combination of femininity *and* sexual maturity, as they result from high levels of female hormones at puberty. This form of curviness also represents a desirable pattern of fat deposition to men because it signifies that women can establish discrete adipose stores to support reproduction, without laying down general fatness for their own use. Throughout the tough years of our species' evolution, men wanted well-provisioned children more than they wanted rotund partners.

The desire for low waist–hip ratios may also be linked to pregnancy, but in a double-edged way. We have seen that women's hips originally became wider to increase the size of the birth canal for our brainy babies to pass through. They then augmented this wider silhouette with external fat, partly because it is a valuable calorie store but probably also because men had come to associate hips with femininity. However, it has been claimed that hip fat has become 'false advertising' of the size of the pelvis beneath. Indeed, the fact that women with broad hips may still experience birthing problems suggests that men can to some extent be 'fooled' by curvy hips.

Conversely, some researchers claim that men's obsession with waist–hip ratios actually stems from a desire to avoid futile dalliances with women who are already pregnant, because waists get wider during pregnancy. I am not convinced by this argument for two reasons. First, we should expect men to be *attracted* to pregnant women so that they maintain the pair-bond throughout their female partner's pregnancies – and indeed they often are. And second, the bulge of pregnancy is a very different shape from that of obesity, or the wide-waistedness of maleness. Indeed, many women's bellies 'bulge forward' during pregnancy, so that their waist–hip ratio is still readily evident from behind – incidentally a sexual position often preferred by expectant couples.

The evidence for the importance of waist–hip ratio is quite good. Many studies show that men prefer narrow waists contrasting with wide hips – the consensus ideal ratio may lie somewhere around 0.70. Although later I will discuss the possibility that there is ethnic variation in this optimum, the general male preference for a low waist–hip ratio has been demonstrated in disparate cultures around the world and brain scan data also show how low ratios fire up areas of the male brain associated with attraction. And as if to validate men's predilections, some studies suggest that women with low ratios do indeed experience fewer problems during birth, bear more children, suffer less chronic disease and may even be more intelligent. These data are hotly contested, but it must be admitted that the waist–hip ratio also does what one would expect it to do as women get older: as the power of sexual selection wanes with age, so the waist–hip ratio rises, a change disliked by many middle-aged women because they see it as de-feminizing.

Breasts

If, to men, curviness indicates maturity, femininity, provisioning and a woman's ability to focus resources on offspring rather than herself, then of course breasts must also be important.

There are various measures of breast-curviness, but waist–bust ratio seems as good as any, and once again men prefer low ratios – slim waists and relatively large breasts. Here too, some studies suggest that this male preference is universal across human cultures, although others have disputed this. Certainly, eye-tracking studies, in which men's gaze is electronically recorded as they look at women, suggest that waist–bust ratio is important – when men first view a woman's body, they look most at the breasts and the waist, and less at the hips, genital area or legs (I will discuss my theory about legs in Chapter 9).

In fact, breasts have generated the single most longstanding and polarised argument about the role of men's desire in the evolution of women's bodies. Uniquely among animals, human females' mammary glands are often large and pendulous when they are neither pregnant nor breastfeeding – or if not large, they are usually an appreciable fraction of their full 'lactation size'. To evolutionary biologists, this uniquely human characteristic really does demand explanation, because possessing such large non-lactating breasts is zoologically bizarre. We have already seen that breasts can severely limit women's movement – most would find it painful to run for any length of time without physical support – and large breasts can also cause back pain. There must be something which balances out these disadvantages of large breasts, but what is it?

One group of researchers maintains that humans evolved always-pendulous breasts because they actually serve a useful function – in other words, they were naturally selected over the course of human history because they increased women's chances of rearing their offspring successfully. One suggestion is that the modifications to the shape of the human head – the dramatic reduction in jaw size and the flattening of the face – meant that human infants had trouble breathing during suckling. And so, instead of the infant muzzle protruding towards a flat breast, a protuberant breast grew out to engage with the flat infant face. This is a reasonable idea, but there are other relatively flat-faced primates in existence, such as gibbons and marmosets, and their females do not have permanently swollen mammary glands. And women with very small breasts do not seem to have undue trouble with the mechanics of breastfeeding.

Another functional explanation relates to the fact that human infants are often carried in their mothers' arms or perched on their hips – something uncommon even among primates. Unlike other mammalian infants who cling to their mother, or who follow her around on four legs, human infants may not always find it easy to reach the breast, so maybe the human breast evolved to dangle bralessly down towards babies' mouths and be grasped by their hands. An extension of this theory comes from those who claim that humans went through a semi-aquatic phase during their evolution, and that prominent breasts were a convenient handhold for infants to seize as they bobbed about in the water. However, any theory that involves active grabbing of lactating breasts seems unlikely to me, as breastfeeding women find this sort of thing extremely uncomfortable.

The evidence for these functional explanations of human breast morphology does not seem convincing to me. Breasts are mechanically very weak, comprised mainly of soft fat with a flimsy connective tissue capsule which gradually deteriorates under the force of gravity. They are hardly structures suited to a mechanical function. Besides, to reach down to and slot into a distant flat-faced infant's mouth, breasts could just as well be long and thin, not globular, and would then impede women's athleticism less. Finally, the function-based theories cannot explain why the breasts swell early on during puberty, before girls can actually become pregnant, and *long* before they usually do.

An opposing group of researchers claims that, whatever function female breast enlargement once had, most of the pendulousness is the result of sexual selection by men. And indeed, it seems clear why men should desire large breasts, even if this conflicts with their potential mate's ability to run around. Breasts are visible stores of fatty calories in their own right, and they also represent an important component of women's general curviness. They are particularly large during pregnancy and lactation – times when women cannot conceive, but when it is crucial for men to maintain their enthusiasm for the pair-bond. Some studies even suggest that breasts swell and become more attractive around ovulation, as a direct lure for human males. Indeed, breast size may directly stimulate provisioning by the male brain – and studies show that men are more likely to take a large-breasted woman out to dinner on their first date. One theory, perhaps unlikely, even suggests that human breasts evolved to maintain men's copulatory interest by mimicking the buttocks they had been so used to ogling before human coitus switched to being a more face-to-face activity.

If always-pendulous human breasts are indeed a sexually selected characteristic, then they probably send mixed messages about age. We have already seen that men tend to seek relatively youthful partners, yet young women tend to have smaller, and certainly less pendulous, breasts. Also, women with smaller breasts are likely to be pleasantly surprised in middle age when their breast shape remains more youthful than their more buxom friends. So as women get older, men potentially face a sexual dilemma between choosing partners with large breasts or with youthful-looking breasts.

Despite the complications related to age, there is strong circumstantial evidence that breast shape is a sexually selected trait. For example, enlarged breasts are permanent features after puberty, they are sexually sensitive only in humans, they are attractive to males only in humans, and their size correlates poorly with their biological function of producing milk. Moreover, breast size is extremely variable between women, and variability is also often a product of sexual selection – some women have breasts ten times larger than others, but the same is not true of eyeballs, thighbones or other functionally selected bits and anatomy. Human breasts even have an additional tell-tale visual advertisement for men – the areola, the pigmented region of skin around the nipple. The areola is a distinctive human innovation, and studies show that it particularly attracts men's gaze, and that they find it especially attractive if it contrasts strongly with the surrounding skin.

So if most of the evidence suggests that human breasts are primarily products of sexual, not natural, selection, women can thank or curse generations of breast-obsessed men, and the generations of women who loved them, for their current shape.

And as it happens, men may derive some surprisingly detailed information from breasts, which could explain why they like looking at them so much – *analysing* them, almost. Symmetry seems to be even more important for breasts than it is for faces. Yes, many men like large breasts, but they like symmetrical breasts more – and small breasts are more likely to be symmetrical than large ones. We assume that breast symmetry indicates genetic health and developmental 'stability', and indeed it does predict future fertility, it reflects high oestrogen levels, and it can be inherited in the form of symmetrically breasted female offspring. There are also suggestions that women with symmetrical breasts are more intelligent. So heterosexual men's apparent obsession with breasts may actually reflect a benevolent wish to optimise his future children's intellect.

While some of these ratios, symmetries and pendulosities may seem abstract, it is clear that even if men do not measure them directly, the womanly curvaceousness these features represent certainly attracts them. The power of curves over men – and their implicit guarantee of maturity, femininity, health and fertility – is clear. As for specific numbers, here are the ratios which one study suggested are preferred by the two sexes when assessing women's bodies:

	body mass index	waist–hip ratio	waist–bust ratio
men's preference	18.8	0.73	0.69
women's preference	18.9	0.70	0.67

It is obvious that the men in the study liked moderately slim women with narrow waists, broad hips and large breasts, but even more noticeable is how similar women's female body ideals were to men's. Indeed, the women preferred slightly more busty and wasp-waisted women than the men did, and certainly a more curvaceous shape than the female participants themselves possessed, so it seems that men are not the only ones imposing expectations of curviness on women.

However, although the sway which the curvaceous female form holds over men's sexual desire seems to have been established, there is one issue which few of the studies have addressed directly. Science tends to focus on averages – the average waist–hip ratio desired by men, for example – but what of variety? Of course, heterosexual men vary in their body-tastes, and this 'between-man' variation can be measured. For example, the recent omnipresence of internet technology means that men who desire female bodies at the extremes of the spectrum can now form what are rather charmingly described as online 'communities', through which they can validate their own preferences for large, small, muscular or pregnant women in the comradeship of other men with similar penchants (and swap pictures, of course) – and also unwittingly constitute an enticing resource for researchers. Individual men's preferences also change according to their experiences – for example, studies show that stressed men prefer women with higher body mass indices, and long-term relationships tend to lead men to develop over-optimistic ideas of their partners' objective attractiveness.

Beyond this between-male variation, there is an even more important, but less easily studied effect. Put simply, individual men do not blindly seek only women who accord precisely with their own preferred ideal physique. Most men find a wide range of female body shapes attractive, so long as they look obviously female, and also somehow 'right' for each individual woman. Maybe this male acceptance of female variability is why ratios matter more to them than absolute measurements. Thus there is also 'within-man' variation, which I like to call 'body-flexibility'. And I believe that men are more body-flexible than many people realise, and often more body-flexible than women.

Women pictured in men's magazines, for example, often reflect a wider range of body shapes than those in women's magazines, and certainly a wider range than is seen in fashion models. For example, studies show that even *Playboy* playmates vary in shape, are not unusually slim and are not becoming progressively slimmer over the years, although they do often have more curvaceous ratios than the average woman. The movie and pop stars who are most lusted-after by men – the real icons – are again a varied bunch: Marilyn Monroe, Jayne Mansfield, Audrey Hepburn, Brigitte Bardot, Debbie Harry, Angelina Jolie, Gwyneth Paltrow, Beyoncé Knowles. Fashion models, in contrast, are generally thinner, less variable and less curvaceous than most women, and often have waist–hip ratios higher than the average teenage boy – but they work in an industry in which heterosexual men may wield less influence than elsewhere.

Thus women may have men to thank not only for their distinctive body shape, but also for the dramatic *variation* in that body shape. And biologically, body-flexibility makes a lot of sense for heterosexual men. They should seek partners who are obviously female, mature, healthy and fertile, but they must not be so blinded by absolute size that they ignore brains, faces, smells, and of course mutual attraction. And once they have found their partner, it makes even more sense for them to be body-flexible – to find her attractive when her weight inevitably fluctuates, when she is pregnant, when she is breastfeeding, when she ages. Investing in a family is a long-term commitment for a human male, so he must be able to adapt to his partner's continually changing body as the years pass.

So when a man says he has not noticed whether his female partner has put on weight, he may simply be telling the truth, and demonstrating a typically male insensitivity to *absolute* body size inculcated by millions of years of evolution.

In this chapter I hope I have convinced you that men seek profound and important information when they look at women, and that over the generations this has relentlessly sculpted women's bodies into their present form. However, this leads us to a paradox which lies at the heart of this book. If male desire is so biologically important, why do women think about it so little?

There are few studies which have sought to discover why women wish to look attractive, but they and the anecdotal evidence tell the same story. I have asked many women why they want to look attractive, and the usual answers boil down to wanting to maintain their self-esteem, or wanting to impress other women. Whether or not those two things are the same I will consider later, but both seem more important than the need to impress men. Of course, these responses may reflect the non-confidential and male-originated nature of my enquiries, but there are surveys which back up my findings. They suggest that for women, when it comes to looking good, impressing other women and maintaining self-esteem are *each individually* at least twice as important as impressing men.

On the surface, this does not make biological sense. We know what men want, we know how this has affected women's evolution, and we also know that for both women and men finding a sexual partner is one of the most important things in life. And yet, heterosexual women do not seem consciously to be trying to attract men, or at least not very much. I am sure that the urge to impress men changes from time to time in a woman's life, but in general its apparent unimportance is very striking.

Although it took this whole chapter to discover what men want, it will take the rest of the book to find out why, as far as women are concerned, it does not seem to matter.

FIVE

Trapped in a vessel of flesh

It is amazing how complete is the delusion that beauty is goodness.

The Kreutzer Sonata, Leo Tolstoy, 1889

'I've always felt very detached from my body. I've always battled against it. It's like I've never felt it belonged to me because I've fought it so much.'

Anonymous interviewee 'D'

We now know how men assess women's bodies, as they watch them from the outside, but what is it like to inhabit one of those bodies?

Over half of the human brains in existence just happen to have found themselves lodged inside a female body, but does this accident of birth make those brains think differently – feel differently? And does being inside a woman's body *make* someone a woman? Awareness of being in a body – one's own body – is a very special form of sensation, but one usually taken for granted. It separates us from the outside world and the outside people. It creates a boundary between 'inside' and 'outside' and that boundary defines an inhabitable volume, a shape of internal, owned flesh. That shape obviously differs between the sexes, but it also varies a great deal between individual women.

This sense of embodiment is important for our study of human bodies because it has been claimed to underlie the acquisition of consciousness and free will in primates. However, although the brain seems particularly dominant in our own species, its self-awareness can only ever be expressed via the body. Descartes claimed 'I think therefore I am' and many of us still believe that the true 'us' resides in our brains. However, each of those brains is imprisoned inside a bony box, borne on top of a fleshy vehicle essential for perception, action, and reaching out to the other imprisoned brains around us.

The mind simply cannot escape the body – it has no choice. And women have distinctive and diverse bodies, so my aim in this chapter is to find out how this distinctiveness and diversity changes the way those minds think.

Is it possible to discover how infants think? They tend not to converse much, and they are distinctly unwilling to discuss concepts like self-awareness and embodiment. However, they often participate enthusiastically and without preconceptions in psychological experiments.

Children start to perceive faces and bodies at different ages, and it seems they also perceive them differently. We start to respond to faces very early in life – within our first few months – but recognition of our own face comes later, perhaps around twenty months of age. Maybe this is not surprising, because for most of human history no one ever saw their own face, so presumably there was little evolutionary pressure on infants to be able to recognise it. Judging from experiments in which infants are presented with images of real bodies and 'scrambled' bodies with bits in the wrong place, an understanding of the topography of the body develops later still – possibly as late as thirty months. Even beyond this age, children still create surprising distortions when drawing bodies, and famously draw heads much too large for those bodies. These differences between the perception of faces and bodies suggest that the two are essentially different, distinct mental processes.

Toddlers soon develop a sense of self – by thirty months of age they know what sex they are and how the sexes differ, and can exhibit shame and pride, as well as empathy for others. By the age of three, children already prefer friends who are attractive and who are not obese, and once they realise that fatness can be controlled they become even more negative about it. By the age of six many girls identify elements of their appearance they would like to change, although this may reflect a realisation that appearance can be modified rather than any profound dissatisfaction with their bodies. However, their sense of self and non-self is already clear and strong. I still remember my youngest daughter, Rose, returning from nursery to recount the distressing story of how her friend injured her knee, and how after her graphic description of the incident, she reassured me by saying, '. . . but it didn't hurt *to me.*'

So what makes us feel, from such a young age, that we are inside our bodies, and distanced from everyone else's?

In recent years we have challenged the old philosophical idea of the brain as master-controller and the body as its passive tool. Philosophers, psychologists, neuroscientists and, most of all, experts in artificial intelligence, have proposed instead that perception, intent, action and intelligence are a product of brain and body working together as an inseparable duo. Some even describe intelligence as being distributed throughout our entire body rather than focused in our head, and this of course raises the possibility that having different bodies necessarily means thinking different thoughts.

At first this may seem a strange idea, but there is a lot to be said for it. For example, the way we perceive the world is entirely dependent on the nature and acuity of our bodily sensations, and we know these differ between individuals. Women have a more acute sense of smell than men and many men cannot discern red and green; there is absolutely nothing the brain can do to eliminate these differences. Also, all sensory information is processed in the context of where the body exists in space, and how it is moving. For example, women are generally shorter than men so they see the world from a different viewpoint. Similarly, we must assume that something touching a woman's thigh feels different from something touching a man's thigh, due to the differing mechanical properties of muscle, fat, skin and hair.

Even more important is the role of the body in controlling

movement. The reason we can hold objects so easily is that our fingers deform when we seize them. The brain 'knows' how squidgy fingers are, but it has no control over that squidginess. Thus the ability to manipulate objects may exist as much in the hands as in the head, as you will discover if you attempt to peel a tangerine wearing metal gauntlets. This is even more true for walking and running, because these activities involve processes of elasticity, deformation, energy transfer and limb-swinging which are not only intrinsic to the limbs themselves, but also too rapid for the brain to keep up with. The brain may be a manager, but it is not a micro-manager – much of the control of movement is 'built into' the body, and not driven by the mind. And women's limbs are different lengths, shapes and sizes to those of men, and each woman's body has its own distinctive arrangement of pivots, weights, pendulums and biomechanical resonances, so to a great extent women simply *have* to move the way they do – their body architecture dictates it.

Studies of this 'embodied intelligence' are continuing apace, and it now appears that even abstract forms of thinking – such as mathematics and language – may also be remarkably dependent on the presence of the body. It is hard to say where all this research will lead, but it certainly suggests that the mind's workings depend on the body in which it finds itself. And women's bodies are very different from men's.

How the mind actually 'finds' itself in that body has received a great deal of neuroscientific attention in recent years. It certainly seems that the perception of one's own body is a very different process from the perception of others' bodies, and this probably explains why women's assessment of their own shape can be so at odds with how other people see them.

Unlike most mammals, vision is humans' most important sense, so it is no surprise that it dominates our perception of bodies, and indeed it is usually the *only* means we have to perceive most other people's bodies. It is for this reason that blind people never develop as accurate a map of the shape of the human body as people who can see. This human visual bias probably also explains why most people 'feel' that their mind, their consciousness, floats just behind their eyeballs.

Magnetic resonance imaging studies of the brain demonstrate some remarkable features of visual body perception. There are two particular brain regions – the 'extrastriate body area' and 'fusiform body area' which become especially active when we see images of bodies, even if those images have all facial features erased. These brain areas react particularly strongly if bodies are moving or making emotional gestures, which probably explains why we find human movement and posture such important contributors to other people's appearance, emotional relevance and sexual allure, and why people like to dance, stroll and play sport together. But one thing is certain: the brain regions which identify bodies are not the same as those which identify faces.

Vision is an exception among the senses because we routinely use it to perceive not only other people's bodies, but also our own – yet the distinction between self and non-self is embedded deep in our image-processing circuitry. Brain scans show that, although there is some overlap, the brain areas activated by looking at our own bodies are different from those activated by looking at other people's. Ingenious experimental trickery also shows that various brain regions may be either more or less activated when the same body is viewed from the perspective of its owner or when it is seen from the viewpoint of a spectator. The top and underside of the feet, the front and back of the thighs – the brain perceives these areas in fundamentally different ways, depending on whether they are a body surface which we usually see on our own bodies.

Brain scanning studies have now progressed to the point at which we can examine more elaborate aspects of the visualisation of our own bodies, such as alterations in body shape. For example, there are particular regions of the brain which flare into activity when people are shown images of their own bodies, computer-distorted to give them larger or smaller waists. However, there are clear differences between the sexes – men exhibit activity in the regions at the back of the cerebral cortex which process visual information about movement of objects in space, whereas women show increased blood flow in areas at the front of the brain responsible for higher-level cognitive functions, as well as deeper regions involved in emotion and fear. It is remarkable that the sexes should differ in such a radical way when it comes to perceiving change in their own body shape, but it certainly chimes with my basic argument that women think about their bodies in different ways from men.

And in the modern world, women's visual self-perception has taken on new, entirely unnatural forms, for which evolution presumably never prepared them. Until a few centuries ago, apart from occasional reflective glimmers in still water, all women ever saw of themselves was a viewed-from-above, foreshortened view of their own bodies – a kind of headless, widened distortion of the visual appearance they presented to others. Look down at yourself now and that view is what you see. Women viewed themselves from within, while others viewed them from without – and never the twain did meet.

However, all that changed with the invention of the high-quality tin-mercury alloy plane glass mirror – probably in Venice in the fifteenth century. For the first time in history women were exposed to an undistorted external view of their own bodies, and it changed their perspective completely. Understanding your own reflection is not innate, but must be learnt – 'feral' children raised by animals do not seem to recognise their own reflection. Even today we know that mirrors do strange things to people. Studies show that exposure to a mirror increases women's self-consciousness, even though they know it actually makes no difference to their appearance. Particular culprits are the concave magnifying mirrors in brightly lit hotel bathrooms, ideal for identifying every last imperfection. Mirrors also alter people's moods – people who are angry can be made angrier by putting them in front of a mirror, and mirrors were strategically located on the Tokyo Metro in an attempt to dissuade people from committing suicide by throwing themselves on the tracks. The presence of a mirror can even make people less likely to lie. It is little wonder that many women have their own 'favourite mirrors'.

And now, photography has distorted the visual self-image even further. A woman can walk away from a mirror, but once a shutter clicks, somebody else possesses a permanent image of their body, fixed at that moment in time – a stationary ghost, almost, but not quite, alive. Many women fear being photographed, and the recent ubiquity of digital cameras and the dumping of their output onto social media have stretched the visual image of the body into even more unnatural territory. Today, many people behave as if events have not truly happened unless they are photographed and uploaded to the internet. Every nightclub attire misjudgement or tentative bikinied foray onto the beach is captured and assessed for suitability for permanent online publication. Many women now use a mobile phone camera rather than a mirror to check their appearance. Indeed, according to recent reports, Karl Lagerfeld's new fashion store has placed tablet computers in its changing rooms so that women can photograph themselves wearing their proposed purchases, and even upload them to social media. Life was simpler when all we saw of ourselves was an inverted decapitated foreshortened torso.

As well as vision, the mind also receives a continual barrage of non-visual information by which it perceives that it is located within the body. Touch is extremely important, and there is a strip of cerebral cortex running right across the top of the brain, the somatosensory cortex, which processes touch information to allow us to literally 'feel' we are inside our bodies. Obviously, we only receive this tactile swell of inside-ness from our own body, and when we get the chance to touch other people's bodies, the experience is entirely different from touching our own. If you touch part of your own body, that body part feels your hand's touch, in a self-affirming reciprocation of oneness. If you touch someone else's body, however, it is a one-way process – you do not get to feel how that other person felt when you touched them. Maybe this distinction explains why we cannot tickle ourselves.

There is much more to 'feel' than skin-touch, however. Our brain also continually receives what is called 'proprioceptive' information from the body – how the limbs are angulated, how much pressure is on the feet, how much tension is in the muscles. Proprioception originally evolved to allow brains to coordinate movement, but it is now also a crucial component of the feeling that you inhabit your own body, as indeed is 'enteroception' – the perception of the presence of your internal organs. This inner sensation explains why pounding hearts and 'butterflies in the stomach' have such profound effects on us. And women's limbs and some of their internal organs are arranged differently from men's, so once again, their bodies must *feel* different from men's.

Another sense which may have an effect on our sense of self is smell, although it has been little studied in this respect. Unless we commute on a crowded subway train we usually have far more opportunity to smell ourselves than to smell others, so smell may indeed help us feel at home in our own bodies. Smell is a strange sense, however – we do not consciously think about it much, our nose rapidly 'gets used' to prolonged exposure to constant smells, yet smell's neural pathways tap directly into our emotions and reminiscences. And of course women often supplement their self-odour with artificial perfumes, to which they can become extremely attached.

Thus almost every sense helps to reassure the mind that it inhabits one particular body, and evidence even suggests that the brain is not too worried about getting these senses confused. For example, we know that one sense can reinforce or even distort the perceptions of another. In one remarkable study, people who felt their own face being touched at precisely the same time that they saw someone else's face being touched, subsequently changed their recollection of what their own face *looked* like. When asked to select from computer-generated blended images of their face and the other person's, the simultaneous touching made them think they actually looked more like the other person. Touch changed the way they saw themselves.

Aside from the senses, there is one further mechanism which makes us feel as if we inhabit our own body. Just as we have a mental recollection of what our face looks like, we also possess a separate recollection of how our body looks and feels – a model of it. To some extent, this model exists to prevent us becoming disoriented – we do not, for example, lose our sense of embodiment just because the room is dark and we cannot see ourselves. Equally, when we dream and cannot 'feel' our body, we remain perfectly capable of generating an avatar-body in which we can play out our dreamy activities.

However, the body-model also has some complex and unnerving properties. First of all, it can be distorted and inaccurate. For example, right-handed people often believe their right arm feels larger than it really is and that their left arm feels smaller. Also, because we cannot see all of our own bodies, we apparently use our visual experience of other people's bodies and extrapolate it back onto ourselves to 'fill in the gaps' which we cannot see. This of course could mean that elements of our body-model end up being just plain wrong. Maybe, for example, I am wrong to assume my bottom looks the same as a youthful Brad Pitt's. And suddenly being confronted with a disparity between the internal model of our bodies and what the real thing looks like can be extremely disconcerting – and as we have seen, women's reactions to such a disparity may be more profound and emotional than men's.

All of these processes – visual and non-visual sensation, and the maintenance of a 'body model' – cooperate to make us feel we inhabit our bodies. However, there is one more component of embodiment which is especially important in making us *care* about our bodies, something which gives them emotional salience, and this is the sense that we own them. It may seem ridiculous to speak of people's ownership of their own bodies, but owning a body 'feels' different from just inhabiting it, or being responsible for its actions. Indeed, women usually have strong opinions about who gets to control their bodies – especially in societies in which they feel controlled, objectified and commodified.

Remarkably, neuroscientists even seem to have found the neuroanatomical location of body-ownership. Experiments designed to stimulate a sense of body-ownership induce activity in the insula or 'island', an area of cortex buried in the deepest fold on each side of the brain. Also, people who have suffered injuries to their insula may disown or neglect parts of their bodies. The insula may also be involved in depersonalisation disorder, in which sufferers feel as if they are passive, disengaged observers of a world which carries no emotional importance for them. The links between body-ownership, emotion and selfhood are being studied in other contexts too. One extreme example is Cotard's syndrome, in which people believe that their body does not exist, and often assume that they are dead, or decomposing. We take ownership of our bodies for granted, and rarely appreciate how central it is to our existence.

The term 'body image' was coined in 1935 and has clung tenaciously to us ever since. I have carefully avoided it so far in this book, but can resist it no longer. A major reason why women's body shapes are so fascinating is that they have two different manifestations – not only do they exist physically, in their own right, but they also exist as a projected image inside the mind, a model of what women think they look like. And that model is more than just a dry, pragmatic estimate of the body's proportions and appearance – it comes loaded with emotional power; it changes women's behaviour, thoughts and fears; it can never be ignored; and surprisingly often it veers away from the objective truth. Also, body image is a fluid thing: it changes over time, it varies between societies and ethnicities, and as we will see throughout the rest of this book, it is malleable, changeable by a wide variety of external influences. But what *is* body image?

I think body image has three essential components. The first of these I will call 'body-perception' – the initial generation of the body-model. Really, this entails many of the things I have just discussed – a sense of self, body-ownership, adapting the brain to 'fit' the body it inhabits – all of these contribute to the body model. However, although you might think that generating this initial model of your body is straightforward, it actually turns out to be remarkably subjective. Just as the shape of a body sets the context for the mind, the foibles of that mind strongly affect how it views its own body.

A person's body-perception is profoundly affected by their personality. 'Personality' may seem a rather unscientific concept, but there is currently a great deal of research going on into personality – partly because personality predisposes each of us to our distinctive behaviours, thoughts and emotions, and partly because evolution seems to have created a wide variety of personalities within animal and human populations. We do not know why evolution makes varied populations, in which different people are pre-wired to react to the world in such different ways, but they certainly react to their bodies in different ways.

Psychologists have developed objective tests which dissect the components of people's personalities, and they have also identified traits which lead to negative or distressing self-perceptions. As it turns out, personality components which involve emotional reactivity and self-consciousness seem to predispose people to negative body-perception, as do perfectionism, self-focusing, acceptance of societal ideals and comparison with others. This list may not come as a surprise, but it is intriguing just how much personality traits affect the initial development of body-perception, even before any subsequent value judgements and emotions have been layered on top. Apparently, no one can ever be entirely subjective and analytical about her own body shape, nor can she develop an image of it without reference to others.

The power of personality in body-perception is also important because, although personalities vary so much between people, an individual's personality is remarkably consistent over time. This explains why studies show that people tend to carry their own unshakable personality-based distortions of their body image throughout their lives. And because personality is a partly inherited, genetic trait, it now appears that body-perception is also partially inherited. Studies of female identical twins demonstrate that genes are more important than upbringing, environment or body weight in determining body image. Also, the relative influence of genes actually increases over the course of adolescence, despite the fact that teenagers might feel that they have more 'free will' to formulate their own opinions by the end of their teenage years.

The second major component of body image is body-satisfaction, or body-dissatisfaction: the subjective contentment or discontent that people feel about their bodies. Of the two opposing terms, I prefer 'body-dissatisfaction' because it implies that satisfaction is the norm – although as we will see, I may be wrong in that optimistic assumption. For example, in one survey, approximately half of all women said they believed that no woman is ever entirely satisfied with her appearance – a belief with which some psychologists would agree.

The most striking feature of body-dissatisfaction is that it changes throughout life, and different people often show similar changes at particular life-stages. Although some researchers claim that female body dissatisfaction can be detected as early as the age of two, many think it first manifests itself between six and eight. At that age, girls especially, start to make value-loaded assessments of their body shape – some years *after* they start to think about the possibility of changing their appearance.

Before the age of ten, girls already talk about fatness much more than boys (boys, if they do talk about bodies, talk more about muscularity). Also, girls exhibit stronger responses to images of fatness or thinness than boys do to images of muscularity or weediness. There is considerable evidence that children between six and ten have already internalised societal standards of what is attractive and what is not, and we believe they use this information to construct 'body ideals' with which they can compare themselves. This may be why they now start to experience body-dissatisfaction: disquiet about the discrepancy between those ideals and their own self-perception.

Early in the second decade of life, girls are already more interested in their own appearance than boys are. At this age, their linguistic abilities are blossoming and language is taking on a new and bewilderingly complex social role. One element of this is that girls develop a discourse about body-dissatisfaction and the desire to lose weight, possibly in imitation of the adult women around them. 'Fat-talk' is a well characterised element of human social communication, and is particularly prevalent among women and girls from the age of ten onwards. While it may represent no more than an expression of girls' body-dissatisfaction, it is likely that girls also use it to cement their place within their social group. Thus fatness is, perhaps, a common interest by which girls can express their self-esteem, while diplomatically emphasising that they do not feel 'superior' to their friends.

Body dissatisfaction becomes progressively more evident during the teenage years. All teenagers must cope with the physical changes of puberty, but while puberty drives boys towards the taller, angular, muscular appearance they often prize, it is a more ambiguous process for girls. Female puberty involves deposition of fat – something all too often associated with greed, unattractiveness and weak will – in distinctively female, unconcealable places. Girls do want to grow up, but the curves that come with maturity can make them feel uneasy. Indeed, in the lead-up to puberty, girls' fat worries are often not focused so much on getting fat *now*, but instead they represent anticipatory concern about getting fat *later*, during puberty.

One US study suggested that by the end of adolescence, 70 per cent of young women want to be thinner, and most of those believe that thinness will make them happier. Increased body-dissatisfaction can be detected in the heaviest 50 per cent of teenage girls, whereas it is only evident in the heaviest 25 per cent of boys. Girls simply appear to worry more about weight. This could be an inherent feature of the female mind – conflicted by finding itself in a body whose sexuality is advertised by adipose tissue – but there are probably other influences at work too. For example, parents direct more body-related criticism (negative *and* constructive) towards daughters than sons, and studies show that daughters perceive that their bodies are discussed more than their brothers'. Also mother–daughter conversations mention bodies far more often than other family dialogues do – and indeed, this could be an important part of the mother–daughter bond.

Beyond adolescence, studies show that body-dissatisfaction is relentlessly increasing among young women. More than 60 per cent are uncomfortable about looking at themselves naked in the mirror, and more than 50 per cent would alter their breasts if they could. More women than men feel overweight, and many more women are on a diet at any one time. Body-dissatisfaction is consistently higher in women than in either men or post-op male-to-female transsexuals. In the US and UK, the most common body-dislikes involve body shape: abdominal fatness is the most common, followed by overall body weight, then lack of muscle tone, followed by cellulite, stretch marks, loose skin and breast size. Quite simply, body-shape discontent is endemic in women in developed countries, and by the end of this book we will have discovered why.

Yet beyond the age of forty, something fascinating happens. Although the body's developmental programme inexorably pulls women's body shapes ever further from the youthful, skinny societal ideal, body-dissatisfaction actually decreases during this phase of life. One study suggested that while 69 per cent of women in their twenties and thirties are dissatisfied about their bodies, that proportion falls to 60 per cent in those in their fifties, and 33 per cent in those over sixty. Clearly, body image cannot be solely dependent on actual bodily appearance. But what else is there?

This brings me to the third and last component of body image, which I call 'body-relevance' – the extent to which people feel that their body-perception, and the body-dissatisfaction derived from it, are important to them as individuals.

Body-relevance is difficult to measure, but we all know it varies a great deal between individual women – some feel that their appearance is emotionally dominant and almost fully defines them, whereas others have a more detached attitude, and seem to consider it largely irrelevant. And it has been suggested that the reason why older women express less body-dissatisfaction is that body shape simply becomes less important to them. Evolutionary biologists might suggest that this is due to a natural, programmed change in the ageing female brain, but sociologists have suggested that as women get older they become less defined by their looks, and more by what they *do* – their roles in society. This may be why older women tend not to choose specific individuals as their role models, and develop more general, more realisable aspirations about how they would like their body to look. We also know that the health implications of body shape become increasingly important to older women, rather than the body's social and sexual functions. Beyond forty, our focus shifts towards wondering how long the body will function, rather than whether it can still impress.

Body-relevance is a very real component of body image, and it seems to be sufficiently flexible to react to changing circumstances. Yet even here the effects of basic perceptions of the body can still exert powerful, and sometimes complex effects. For example, blind women have been shown to be less concerned about their body shape than those who can see – presumably because they have lost the main sensory route by which others perceive their own bodies, the bodies around them, and bodies in the media. And the earlier in life women lose their vision, the lower is their body-dissatisfaction and body-relevance. Also, in a strange reversal of what happens in sighted girls, blind girls' self-esteem is reduced more when their parents suggest they are thin, than when they suggest they are fat.

Age, perception, and personality all inform women's sense of how important their body shape is to them. Of the three components of body image, body-relevance is the hardest to define and hardest to measure, but it may prove to be the easiest to change in women who want to escape the prison of their own body image.

It is impossible to overstate the power of body image. It affects everything we do and everything we feel – and it is hard to think of anything more powerful than that. It has us in its thrall. In fact, every person behaves differently *because* we each possess different bodies and different body images. We have already seen how the mind affects the image of the body, but how does the image of the body affect the mind?

The simplest body-influenced behaviours are, unsurprisingly, behaviours directed at tending to the body itself, or our image of it. The average woman spends a large proportion of her waking hours in body-tending activities, and over time these can become increasingly complex, personalised and protracted. There are four main body-tending behaviours, and women switch seamlessly between them at any time. The first is 'checking' – assessment of one's own appearance in various ways. Women look at themselves and peruse their reflection, but checking also entails touching, squeezing, smelling, measuring and manipulating – sucking in, puffing out, changing posture, even jumping up and down. Checking can be stressful, and is usually confined to short bursts lasting no longer than two minutes.

The second behaviour is 'fixing' – short-term correction or camouflage of perceived bodily imperfections with adjustments to makeup, hair or clothes (I will discuss these in much greater detail in Chapter 9). Fixing is often followed by a renewed bout of checking.

The third behaviour is 'coping', or developing strategies to mitigate problems which cannot be 'fixed'. Coping may involve compensating for defects by emphasising another feature instead, or simply receiving reassuring, validating, affirming feedback from a friend or partner. Coping seems to be all about modifying the body-*relevance* of a particular element of a woman's appearance, rather than changing her satisfaction with it. This is why women ask their partners' opinion about their appearance – and many partners seem implicitly to understand this, and so do not express their real opinions. Of course men check, fix and cope too, but they spend less time doing it. This could be because they usually have higher body-satisfaction in the first place, or because they analyse their bodies in less complex ways, or because they have a narrower repertoire of fixing and coping strategies available to them.

It is the fourth body-tending behaviour which can be the most destructive. 'Avoiding' is what women do when they cannot find a solution to a particular body-image problem – no matter how much they try to fix and cope, the perceived defect cannot be corrected, hidden or ignored. Instead they restructure their life to avoid exposing that defect to the eyes of the world, and in some cases this can become socially crippling. Some 'avoiding' behaviours seem relatively trivial – many women never weigh themselves, and one study suggests that one-quarter of all women in the UK do not own a full-length mirror because they do not want to see a reflection of their entire body. However, these common avoiding strategies represent the more benign end of the spectrum of severity. Many women always wear baggy clothes, for example; some will only undress in the dark, even when alone; others refuse to go swimming, or to a beach, or to a communal changing room; some avoid dating, or sex, or any physical contact at all; and some rarely leave the house. Women often keep these behaviours secret because they worry that no one would understand them. And sometimes they have friends or admirers who would *like* to see their bodies, or touch them, but cannot. Everywhere, every day, body image places restrictions on women's lives.

Surveys suggest that poor body image adversely affects the sex-lives of at least one-third of all women, and one-third believe that losing weight would improve their sex-lives. Low body image makes many women feel negative about sex, and makes them avoid certain sexual positions, or situations – for example, being on top of their partner, or having sex with the lights on. Because prevailing stereotypes tell women that their main sexual role is to visually arouse men, body-dissatisfaction often distracts them emotionally from the act of sex, and the enjoyment of sex, and that distraction can lead to sexual dysfunction. For all sorts of reasons, it would probably be best if people were able to forget what they look like when they are having sex.

Indeed, body image controls almost every aspect of romantic relationships. Women who perceive themselves to be more attractive are more selective about choosing their partners. They are also especially picky about the characteristics which they feel are most attractive in themselves – they seek men whom they think are 'similarly optimal' in body shape, athleticism and so on. They also tend to select men who have faces which are more symmetrical and more masculine. Other studies show that women with 'feminine' lower waist–hip ratios show the same tendency to pickiness. On average, heavier women describe their romantic and sexual relationships as being worse, or more likely to end, they believe their partners consider them less warm or trustworthy, and they form relationships with men whom independent observers perceive to be less desirable. Alarmingly, women with poor body image are also less likely to use barrier contraception (whereas it is men with *high* body image who are less likely to use it).

Women show an extremely strong correlation between body image and self-esteem, and this is evident from the start of puberty, and probably earlier. Surveys of teenage girls show that they believe that feelings about their appearance control their self-esteem and positivity, rather than the other way round. In adulthood, other studies suggest that body image continues to be the most important determinant of women's self-esteem, and it affects how they deal with others in complex ways. For example, women who perceive themselves to be attractive are more likely to become angry, and to use anger as a tool to get what they want or make other people treat them better.

Considering its undoubted power, one of the strangest features of body image is that it is often an inaccurate portrayal of women's bodies and how attractive others perceive them to be. Several studies suggest that women's assessments of their own bodies do not match the assessment of independent observers. Body image is often just plain *wrong*, despite the fact that the mind spends so much time and energy formulating it. For example, women with a good body image (no matter what kind of body they have) seem to show little day-to-day variation in that body image, whereas women with poor body image exhibit a great deal of fluctuation – is it really likely that the two groups differ so dramatically in how much their appearance changes?

Body image distorts and deceives, and it even distorts what women see in others. Studies with computer-morphed images of celebrities show that women with poor body image consistently underestimate the weight of female celebrities, whereas women with a good body image are far more accurate. Also, eye-tracking studies show that women with poor body image find it harder to avert their gaze from pictures of women who are relatively light or heavy.

Another strange feature of body image is that women's perceptions and body-ideals seem remarkably easy to modify. If women are twice asked to select images of female bodies which they think are the 'ideal' shape – once before being shown pictures of skinny women, and once after – then their ideal shape becomes smaller following exposure to the images of skinniness. The same is true if women are asked to computer-morph pictures of their *own* bodies to their preferred size – they want to 'shrink' themselves more if they have recently been exposed to pictures of slim women. The brain, it seems, cannot help distorting the bodies it sees, and then elevating that distorted image to the status of something to aspire to, something to be longed for.

'I am therefore I think.' It is wrong to think the mind can separate itself from the body. The body is essential not only for allowing the mind to perceive and influence the world, but it is also central to the mind's sense of self, and its engagement in that world. We feel that our selves exist inside our bodies, that they 'own' them, and we also each create a body image which we can assess and alter, love or despise. Because of this, we each think differently because of the body shape we happen to possess. The body is not a passive vessel – quite the opposite: we simply cannot stop it controlling how we think.

However, if body and mind are so intertwined, does this not mean that – to return to that idea from Tolstoy with which we began the chapter – beauty *is*, to some extent, goodness?

Certainly, in most cultures this is tacitly assumed to be the case. From an early age we are told stories of good characters who are pretty and slim, and bad characters who are ugly, deformed or fat. And whatever we do, we cannot shake off this beauty-slimness-goodness chauvinism. Studies show that adults expect attractive children to be better behaved, better adjusted and more responsive to discipline and constructive criticism, even though there is no evidence that this is the case. So it comes as no surprise that children start to make the same assumptions as early as the age of six, and that this beauty-bias continues throughout life. Controlled experiments show that people believe attractive adults are more intelligent, sociable, healthy, dominant and sexually experienced – as jurors they are more likely to acquit an attractive person of a crime, and in job interviews they are more likely to offer them employment. One US study showed that waitresses' tips are affected less by the standard of service they provide than by their youth, slimness, blondeness or breast size.

So despite its inaccuracy and subjectivity, I believe this is why body image is so all-controlling – it is an internal acceptance that people will judge us on the basis of how we look. And this is especially true for women because, as we will see later, body shape is of particular social importance to them. Body image is not logical or objective – it is only loosely anchored in our real appearance and is skewed by emotion, personality and the random chance of experience. It panders to the unfairness, chauvinism and discrimination of others. It skews beliefs, behaviours and self-esteem to the point at which some people can no longer live healthy, happy lives. Yet we simply cannot help it.

SIX

Comfort and discomfort eating

And the Lord God said unto the woman, What is this that thou hast done? And the woman said, The serpent beguiled me, and I did eat.

Genesis 3:13

'I was controlling how many calories I ate each day, and I remember on one occasion I ate two cream cakes and just felt so dreadful that I took some sleeping pills to make sure I didn't eat anything else that day. It must have been the afternoon, and I must have assumed that my children would be all right.'

Anonymous interviewee 'E'

In the developed world, we exist within a cult of thinness, and that cult is focused on women's bodies.

We are continually surrounded by messages, some subliminal and some less so, that being thin is a good thing – it equates with an almost moral goodness. We are told that being fat is bad for you, personally as an individual, but we are also all encouraged to believe that fatness, and especially female fatness, is an affront to others. Thus women are what they eat: they are personally responsible for the body shape they inflict upon others' gaze. Being anything other than thin is seen variously as unhealthy, lazy, sluggish, lower class, dirty, inadequate. It is considered to represent a weakness of moral character, an inability to manage one's lusts, a pathological failure to accept one's social responsibility to be slim and non-burdensome.

Men can, and do, find ways to escape this cult, but women often face a harsher choice – either live within its rules, or be cast into the outer darkness of un-thinness. As a result, some women effortlessly match the thin ideal, but many others are ensnared in a tangle of shape, food, weight, guilt and comfort-seeking, and that tangle is the focus of this chapter. Of course women can do things to change their body shape and size – they can diet or exercise, use drugs or have surgery – but all too often they still do not get the body they want, or the body others seem to expect of them. You can give up almost all the enjoyable things in your life, but you cannot give up food, and women *need* fat if they are to have a feminine shape, so what are they supposed to do? Why do so many women diet, why do those diets so often fail, and why is this failure perceived as those women's own, personal fault? Why does food frequently make women feel guilty? Why, in short, are shape, food and mood so intertwined?

Many have bemoaned the cult of thinness, and in a later chapter I will investigate when it developed. However, rather than just complaining, I believe that if we wish to challenge it, we must find out why it has proved so powerful. After all, there are good reasons why food, mood and shape are interlinked, and those reasons stem from the way women lived during the vast majority of human history. Sometimes times were good, but sometimes times were hard. Often there was not enough food, and often women and their children died because of it. Yet those women and children who perished were not our ancestors – our forebears were the ones who gorged when food was plentiful, survived thriftily through the famines and lived to gorge another day.

And, paradoxically, that urge to eat – that pressure to fix food, hunger, appetite and eating deep inside our instinctive drives – is where the cult of thinness came from.

That prehistoric fixation on food persists today, and explains why body shape and weight are so inextricably connected to contentment and discontent. Studies suggest that perhaps 87 per cent of women have dieted at some time. More than 70 per cent of teenage girls want to be thinner even though most of them are not overweight. In contrast, less than 10 per cent want to be larger, and many of these girls are, in fact, underweight according to clinical measures (indeed, my youngest 'anonymous interviewee' actually laughed out loud when I suggested that some women wanted to gain weight).

Studies in developed countries show that teenage girls and young women believe that being thinner will make them happier, healthier and prettier, while research across the world shows that heavier girls and women are less satisfied and have lower self-esteem. In approximately one-third of women – a huge proportion – the urge for thinness drives them to attempt more extreme weight-loss strategies, such as crash dieting, fasting, purging or laxatives. And these trends continue throughout life, although older women are more likely to rationalise their weight loss attempts as being part of a programme of general health improvement. Being complimented for slimness usually carries more emotional potency than being complemented for facial beauty, intelligence or success.

Much of this urge for thinness, for smallness, has its roots in the ideas I discussed in the last chapter. Each of us inhabits a body which takes up space and intrudes on the world outside. We also each acquire a personal space around us, within which we stride, gesticulate and comport our bodily affairs, and from which we prefer to exclude most other people. Yet women's bodies usually take up less space than men's, and from an early age they learn that this smallness is desirable and should be emphasised. Boys learn to take long, bold strides, to lope and sway as they move, to swing their arms and sprawl luxuriously on chairs, beds and, presumably, girls. They are, in other words, trained to take up as much space as they damn well want, because that is what men 'do'.

In contrast, girls learn what I would call 'spatial timidity' – they are trained that to be feminine they must take short steps, and tuck themselves into compact little poses which minimise their physical occupancy of the outside world. They realise that small is good, and smaller is better. In some cultures, women also learn to bow, hunch their shoulders and speak quietly to enhance this effect. One of the most elaborate examples of spatial timidity is the way women often practise tricks to appear small on camera – that they should stand slightly side on, with one (flexed) leg hiding the other (straight) leg from the lens – and it is not only actresses on the red carpet who do this. I suspect that spatial timidity is largely learnt – and I know for a fact that some of my female university admissions interviewees are instructed by their schools to cross their ankles and hold their arms to their sides in the belief that such physical compactness will convince me of their aptitude in organic chemistry or literary criticism.

Becoming anything other than thin means, of course, intruding further into the outside world than is deemed acceptable. To make life more complicated, any move away from thinness usually involves the acquisition of fat, and female fat, as we have seen, holds an ambiguous place in our thoughts. It is of course an essential and desired component of the distinctive human female form, yet it is usually viewed negatively when in excess – even though there is no clear definition of what 'excess' is and this can change according to fashion. This double-edged nature of fatness is thrown into stark relief by the fact that, in contrast, such negative connotations are not associated with the other way in which women expand to inhabit more space: by becoming pregnant.

So becoming larger by acquiring fat is a complex psychological process for women. If a man gets fatter, then he is probably seen as slightly less attractive, but if a woman acquires fat, then some people may think she looks *more* attractive – men, for example, or friends and relatives more concerned about her health and wellbeing than whether she conforms to the thin ideal. And of course this validation and encouragement of female fat gain can be immensely confusing – as it conflicts with her prior indoctrination to be as small as possible. Also, unlike men, women have a variety of different locations in which they can store additional fat – thighs, buttocks, bellies, breasts, arms and so on – and people respond differently to fat acquired in those different places. Indeed, many of those fat depots have become the focus of sexual fetishes, in a way that simply does not occur with men's pot bellies.

Other than spatial timidity and attitudes to fat, the third influence which feeds into this tangle is our beliefs about eating – and here again, a simple, innocuous difference between the sexes has contorted into a tangle of assumptions, interpretations and value judgements. Because women are smaller than men, they need less food, so they usually eat less. We are used to seeing men eat more than women, so we have come to believe that eating little, or eating slowly, is feminine. In the media, men are presented as having hearty, healthy appetites for food, whereas women are seen as cautious, picky and restrained in their eating habits. I am reminded of the scene in a US television comedy show in which four males compete to eat enormous steaks, and the loser is told to look for his money in his handbag, next to his tampons.

Physically, women certainly do eat more slowly than men, taking disproportionately smaller bites and sips – smaller than can be explained by the fact that their mouths and stomachs are less capacious. Also, unlike men, women do not eat more when presented with larger food items – when offered an unlimited supply of sandwiches, men will eat a greater total mass of food simply because each individual sandwich is larger. There are obviously some complex brain processes going on here, yet studies consistently show that restrained eating is seen by both sexes to be a feminine trait. Also, women clearly understand the social importance of how much they are *seen* to eat: they eat less when they are in a large group than when they are alone, and they also eat less when they are dining with a desirable man than when they are dining with a less desirable man or, heaven forbid, another woman.

So beliefs about size, fat and appetite have powerful effects on women's attitudes to their shape, but there are also deeply ingrained biological imperatives which mean that food itself has a profound and direct influence on mood.

For any creature to survive it is essential that its motivations and behaviour are controlled by food, hunger, and the opposite of hunger, satiety. Of course all animals are driven to eat when hungry and to cease when sated, but it is more complicated than that. Most animals are also hard-wired to *over-eat*, because this allows them not only to store excess nutrients to see them through times of want, but also frees up time in the future when they need not eat – so they can do other important things, such as breeding. Of course, this tendency to over-eat must be counterbalanced by the instinct to *stop* eating when necessary, so it too is built into our brains – people rarely feel hungry when having sex, for example. And we assume that most of this appetite pre-programming comes to us in our genes, and indeed, studies with identical twins suggest that appetite is a strongly inherited trait in humans.

Appetite is such a life-or-death phenomenon that it comes as no surprise that it is probably the most complex interaction that exists between the body and the brain. For example, pioneering animal experiments conducted over a century ago showed that hunger is partially relieved by mastication and swallowing of food, even if that food does not reach the stomach. Conversely, distension of the stomach in the absence of chewing and swallowing was shown to reduce hunger, although not as much as one might expect. More recently, physiologists have shown how these simple stimuli activate particular brain centres involved in hunger and satiety, as well as other regions associated with heat, cold, sleep and aggression. These other links explain why we eat more when the weather is cold than when it is hot, why a big meal makes us sleepy, and why hunger makes us snappy.

There is also a wide variety of internal chemical cues by which the brain calculates the body's hungriness for calories and other nutrients. For example, in the short term the brain monitors the level of glucose in the blood, as well as the amounts of insulin released by the pancreas. Also, in Chapter 2 I mentioned leptin – the hormone released by adipose tissue which gives the brain a medium-term measure of the amount of fat in the body. In humans, low leptin levels are a sign that energy stores are depleted and that appetite must be increased, yet high levels of leptin do not consistently suppress appetite – suggesting that for most of human evolution our preoccupation was with insufficient rather than abundant food. Unlike many of our relatives humans seem to be a 'famine species', with a physiology adapted to cope with scarcity, and unprepared for plenty.

As more research is done, the ways in which digestion and metabolism control the brain have been shown to be ever more numerous and intricate. Ghrelin, for example is a hormone made by the stomach as well as other organs, which has a direct effect on the brain to increase appetite. In fact, some researchers now believe that there are over one hundred chemicals which may attune eating behaviour to the body's nutrient reserves, the gut's contents, and what the mouth believes it has just gulped down. Many of these chemicals act on the brain to alter our behaviour in ways more unexpected and subtle than we can currently understand. An indication of the profound effects which these chemicals may have on the mind is the rather psychedelic names some of them have been given: endogenous opioids, cocaine-and-amphetamine-regulated transcript, and endocannabinoids.

One way in which eating may exert an almost drug-like effect on mood is by the eating of 'comfort foods'. Most people comfort eat to some extent, effectively self-medicating with foods which they either psychologically associate with calm, secure times in their lives, or which carry the correct cocktail of nutrients to give them the desired 'warm' feeling. Men tend to select hot, meal-like foods as comfort foods, such as meats and stews, whereas women are more likely to select cold or even chilled, snack-type foods, including sweets and ice cream – perhaps they do not wish their dietary indiscretions to look too much like a full-blown meal.

The comfort foods favoured by women are often high in carbohydrate, especially simple sugars, and this seems to be the key to how they work. High-carbohydrate meals reduce cognitive performance and induce sleepiness, but the effects they have on mood depend on people's habitual diets. In people used to meals high in carbohydrate and sugar, they can reduce depression and anxiety, whereas in people not used to them, they can have the opposite effect. There is also evidence that people's personalities have a strong influence over the mood-altering effects of comfort foods. One study showed that in people prone to psychological stress, high-carbohydrate meals reduced the mental and physical effects of experimentally induced stress – anxiety, increased pulse, elevated stress hormones, and reduced cognitive abilities. In contrast, carbohydrates did not exert the same stress-relieving effects on people not usually prone to stress. It has even been suggested that some people use carbohydrate to alter the levels of serotonin in their brains – a similar effect to that achieved by many antidepressant drugs.

Another emotional effect of food is guilt, and here our ambivalent attitude to food and fat becomes very clear – especially for women. Guilt is often associated with particular foods, and indeed advertisers often imply that their food products are guilt-inducing as a badge of just how appetising they are. Studies show that women are much more prone to food-guilt than men – probably because they worry more about how food will affect their body shape. One study in the Netherlands suggested that almost all young women experience food-guilt of some sort, and that this guilt is induced far more effectively by snacking between meals than by the eating of large or calorific main meals. Of course, we have just seen that snacking on 'non-meals' is precisely the way in which women use food for comfort, suggesting that guilt and comfort are an uneasy duo central to women's habitual eating behaviour.

Guilt is a more complex emotion than comfort, however, and it probably involves not just low-level hunger and satiety centres in the brain, but also higher cognitive processes. Because of this, food-guilt is probably partly learnt – built into women by the society in which they live. Traditionally, women are seen as having a greater role in food preparation than men – they are told that it is their role to prepare food for men and children. However, at the same time, societal pressures also tell them that they must not eat too much or put on weight – that they must deprive themselves of the enjoyment of the food they prepare for others. Women's magazines focus mainly on the holy trinity of beauty, sex and food, yet the last of these is presented as an essentially transgressive thing for many women, with the implication that it will destroy their chances of attaining the first two. Food is compulsory, yet prohibited.

If food can affect mood, then just as surely, mood controls what women eat. The mechanisms which underlie this control are probably mainly located in the outer layers of the brain responsible for 'higher' cognitive process, the cerebral cortex, and some claim to have further narrowed their location to an area known as the orbitofrontal cortex. The involvement of these higher regions means that mood can control appetite in complex ways, inducing subtle preferences for particular foods or even portion sizes, and that appetite can change in response to labyrinthine social and cultural mores of eating and food sharing.

This higher cognitive input also means that links between mood and eating can be learnt. In fact, many animals use their early eating experiences to inform their later food choices – from fish to mammals, a primary function of memory and motivation seems to have been to encourage animals to seek out foods that have sustained them in the past. Because eating habits are often learnt, people come to associate foods with particular situations, crunching a salad in the open air on a sunny day, or indulging in junk food after an alcohol-suffused night out. Just like an ex-addict avoiding their former haunts, women can change what they want to eat merely by changing where they go.

One of the most mood-laden learning environments is the childhood home, so it is hardly surprising that girls learn many of their eating habits there. However, studies using hidden cameras suggest that parents often 'mis-train' their young daughters. For example, parents frequently use food to pacify children, especially young children, and to alleviate distress or placate anger which is not itself related to food – a lesson easily remembered into adulthood. Also, although parents may keep 'unhealthy' foods in the house, they deliberately restrict children's access to them to encourage them to eat other, more 'healthy' foods. But this often leads to the forbidden foods seeming even more tempting, and subsequently causes these children to furtively eat more of them when their parents are not present. In fact it seems that children, and girls in particular, learn their eating habits by example, so having parents who eat a varied diet free from guilt and psychological manipulation is probably best.

Many women believe that psychological stress affects what they eat, and that as a result, stress, eating and weight gain constitute a vicious circle conspiring to make them fat and unhappy. However, although stress has very real effects on food intake, those effects are complicated, and sometimes paradoxical. Some data suggest that stress and anger at a particular phase of life correlate with weight gain over the following years (as many as thirteen years in some studies). Long-term stress induces secretion of the hormone cortisol from the adrenal glands, while short-term stress may induce ghrelin production, and both of these may increase appetite. Meal size may not increase, but snacking and 'covert' eating often account for any weight gained – and a woman can acquire considerable weight within a few years by eating only a few extra calories each day.

Yet the science of stress and eating is not simple, and in some studies stress clearly induced profound and consistent *reductions* in appetite – so maybe the female brain reacts differently to different types of stress. For example, there is evidence that boredom, anxiety, excitement, happiness and sadness exert subtly different effects. One interview-based study suggested that women are likely to eat less after sad events in their lives and to eat more after happy events – and that they tend to self-medicate with sugary snacks, perhaps to alter their brain serotonin levels. A different study, of women who experience frequent food cravings, suggested that cravings are specifically induced by boredom and anxiety, rather than stress or lack of food.

The ups and downs of romantic relationships also induce mood-related changes in appetite and body shape, although these changes are difficult to study. We all know people who have gained weight once they were in a settled romantic relationship, and others who have lost weight after a relationship ends, and there are several reasons why this may happen. Contentment itself, lack of stress, and – dare I say it – complacency about one's personal appearance could all increase food intake in happy couples, as could the simple fact that eating together is often one of their main bonding activities. From the evolutionary point of view too, it makes good sense for newly entwined couples to pile on the pounds, as this provides both of them with a calorie-rich adipose buffer against the impending demands of pregnancy, breastfeeding and childcare. Smug-couple chubbiness is the adipose equivalent of a healthy joint bank account.

Across human societies, the socially recognised start of romantic and sexual unions is usually marked ceremonially, by marriage. This may seem an archaic custom, but it is remarkable just how much women's self-esteem, body-dissatisfaction and food-guilt are brought into focus as a wedding approaches. Surveys show that more than half of women want to be slimmer by the time of their wedding, more than half intend to go on a diet, and more than half intend to exercise more. At least one woman in ten is told *by others* that she should lose weight. Many women deliberately buy a wedding dress which is too small for them, in order to force them to lose weight before their wedding. However, despite the antiquity of marriage, women's most commonly cited reason for wanting to be slimmer on their wedding day seems surprisingly modern – it is not to appear more attractive to fellow-revellers, but rather to look good in the wedding photos.

For many women, the most obvious way in which mood affects their food preferences is during the premenstrual phase of their cycle, when many experience urges to eat high-calorie foods, especially chocolate. Chocolate is a complex food with a pleasant taste and texture, is high in fat and sugar, and contains a wide variety of biologically active substances, including the xanthine chemicals caffeine and theobromine, both of which may be responsible for its almost addictive qualities. One US study estimated that 45 per cent of women experience chocolate cravings, and many say that chocolate makes them happy, although the scientific data for this are conflicting, with some studies suggesting that chocolate improves mood, in the short term at least, and others indicating that long-term chocolate eating correlates with depression. Of course, this could be because chocolate is the ultimate guilt-inducing food, or it could indicate that depressed women are more likely to self-medicate with chocolate.

Apart from its psychoactive constituents, chocolate also contains simpler nutrients well known for making food desirable – for example, the body is attuned to liking sugar and fat because they provide large quantities of valuable energy. Indeed, it may be the high calories in chocolate which make women desire it at particular phases of their cycle. Unlike the metabolism of men, the metabolic physiology of women with regular menstrual cycles is in a state of constant upheaval, as levels of metabolically-active hormones soar and plummet. Oestrogens are known to suppress appetite in many animals, and indeed, women eat least around the time of ovulation when their oestrogen levels are at their peak. Oestrogen fluctuations also change the flows of fluid in and out of body tissues so that breasts and bellies *feel* different at different phases of the cycle. Thus women's food intake varies throughout the cycle partly because of direct hormonal effects, and partly because they want to reduce the fluctuations in how fat their body feels. In addition, there is evidence that oestrogens alter the effects of endocannabinoids on appetite, and that levels of serotonin, a brain chemical linked to mood, show regular changes across the menstrual cycle. Thus premenstrual cravings for chocolate could reflect a simple desire for one of the most calorific foods ever created, or a deliberate attempt to manipulate the brain's internal mood-chemistry.

So body shape affects mood, food affects mood, mood affects food, and obviously food eventually affects body shape. Women have inherited complex and powerful control systems to regulate these interrelated phenomena, naturally selected over millions of years to optimise their chances of survival. These control systems did not evolve to make women's lives difficult: they evolved because they were beneficial. However, when women decide to change the body shape their food-mood-shape control system has made them, they discover just how stubborn and unyielding that control system can be.

Dieting is the most common strategy used by women who want to change their body shape or size. In today's 'obesogenic' world many of us simply assume that it is natural, responsible even, to control one's weight by consciously managing one's food intake. Yet until the last few decades humans rarely needed to exert conscious restraint on their eating, simply because there *was* no excess food. In contrast, there now exists a persuasive and highly lucrative dieting industry which promotes the idea of conscious control, especially to women. Particularly concerning is that this industry also promotes an 'ideal' body shape which is probably too small. And as we will see, encouraging moderate-weight women to diet can be harmful.

Women usually diet to lose weight, yet we saw in the last chapter how women's assessment of their supposed excess weight is rarely objective. Unsurprisingly, women who diet frequently usually report a larger discrepancy between their perceived body shape and their 'ideal' body shape. However, picking apart these findings shows that the tendency to diet may have little to do with actually being overweight, or even with having an unreasonably small 'ideal' body shape. Instead, the impetus for dieting is more likely to be a distorted *view* of one's own body shape and size. Dieting also seems to be linked to mood, and in women more than men there is a correlation between depression and dieting, and extreme methods of weight loss in particular.

One notorious feature of women's attempts to lose weight by dieting is that they are surprisingly unsuccessful. Research suggests that only one-third of female dieters stop their diet because it has brought them to the body weight or shape they desired. Diets have a remarkable failure rate, and women often blame this failure on themselves. In fact, studies have shown that women on a diet often consume just as many calories as they did before, and that a significant number actually consume more. In other words, when women take conscious control of their food intake, the weakness of that conscious control in the face of millions of years of natural selection becomes obvious.

First of all, humans seem to have a natural inbuilt tendency to underestimate how much food they eat – not just a devious penchant for lying about their greed, but rather a basic cognitive inability to audit incoming calories. And, of course, we evolved that inability because for most of human history a tendency to underestimate food intake was no disadvantage to us. The second cause of 'diet failure' is that people may assess portion sizes on the basis of volume rather than calorie content – so eating less calorie-dense foods rather than lower food volumes may be a better way to lose weight. The third factor is that once women start to consciously think about food, well, they start to really *think* about it. They think about eating and they think about how much they like eating, and this seems to make it almost impossible to eat less. And fourthly, over the course of a dieting day, eating less at some times means that women become much hungrier before the next meal, making them more likely to 'rebel' against their diet or 'give in' to hunger by snacking – and we have already seen how effectively women can cognitively separate the idea of snacking from the concept of ingesting calories.

It is for these reasons that 'voluntary' dieters who plan their own diet are usually less successful than those who sign up to commercial, 'imposed' dieting regimes (as long as those imposed-dieters stick rigidly to their regime). However, imposed dieting brings its own psychological implications too – for example, women can interpret their success as depressing evidence that they are themselves incapable of controlling their own body size and shape and a fear that they will revert to their old eating habits once left to their own devices. Imposed dieting puts them at the mercy of the very industry that helped them feel so body-dissatisfied in the first place. In addition, the body's food-mood-shape control system is still able to thwart even the most savagely restrained dieter – as a woman's body loses fat, her physiology responds by reducing its metabolic rate and heat production to conserve calories and stymie weight loss.

Even for the minority of women who consistently and successfully lose weight by dieting, the bodily control system still has one more dirty trick up its sleeve. It is a well-recognised phenomenon that periods of dietary restraint are usually followed by compensatory periods of over-eating. This 'post-starvation hyperphagia' is the body's attempt to make good the energy deficit it has incurred, but it is not a perfectly attuned response. Compensatory over-eating is probably caused by the body's control mechanisms detecting that depots of fat and lean tissue have shrunk, and attempting to rectify the situation. However, the period of hyperphagia continues longer than it needs to – longer than the body requires to replenish its reserves – so that many women end up weighing more and having more fat than they did before they dieted.

This 'rebound' fat gain can lead to depression and a sense of helplessness, which may in turn increase appetite, cause weight gain, and precipitate repeated self-defeating cycles of dieting, hyperphagia and body dissatisfaction. What is particularly intriguing is that the risk of weighing more after post-starvation hyperphagia is higher in women who were not clinically overweight in the first place. So controlled diets may be a good idea for women who need to lose weight for health reasons, but they may be *entirely* counterproductive for the majority of women who do not.

Dieting also has other, unintended effects on women's health. Women on a diet tend to make more healthy food choices, and indeed, many do not clearly differentiate between dieting and healthy eating – and this may explain why women usually eat more fruit, vegetables and fibre than men. This could of course be seen as a good thing, but as we will see in the next chapter, it is possible for women to use the disguise of 'healthy eating' to mask risky and even pathological weight-loss strategies. In addition, women may use smoking, or even cocaine, as an adjunct to dieting because these drugs reduce appetite. Indeed, in the US, smokers weigh an average of 3kg less than non-smokers.

Dieting also has profound adverse effects on the brain, increasing the incidence of depression and other mental disorders as diets becomes ever more protracted. Women on a diet show clear global impairment of their higher cognitive functioning, although whether this is due to the brain being starved of nutrients or simply being preoccupied by thoughts of food is not clear. Whatever the cause, dieting effectively makes women (temporarily) less intelligent.

Exercise is the second most common way women try to lose weight and its popularity seems to be on the rise.

Women are more likely than men to opt for exercise as a way to lose weight. Some believe it may lead to selective loss of fat from particular parts of the body. Failing that, exercise usually leads to increased muscle tone, something which can make women feel happier within their bodies, even though the shape and weight of that body may not have changed – a 'felt' improvement, rather than a 'seen' one. While exercise is usually a slower way to lose weight than dieting, it often enhances body-satisfaction faster, increasing women's drive to persevere with their weight-loss programme. And exercise is increasingly considered to be a more socially acceptable and less faddish way of losing weight than dieting – women usually feel more confident openly discussing their own visible, virtuous efforts at the gym than recounting the details of a covert calorie-reduction plan.

However, the extent to which exercise reduces weight varies a great deal between individual women. Just as women vary in their shapes, metabolic strategies and responses to dieting, they also differ in their metabolic responses to exercise. The minority of women who assess their size by weighing themselves are often disappointed that their weight *increases* soon after embarking on an exercise programme, although they may attribute this to the fact that muscle weighs more than fat. But most of the variation in women's responses to exercise is probably due to how exercise alters their appetite. Some women continue to eat the same amount when they start to exercise, and thus lose weight because of the increased drain on calories, but some women eat more. Exercise increases women's drive to eat, so they can fuel all that exertion, but it also increases their satiety responses, so that they may feel full earlier on during meals. Whichever of those effects predominates – increased appetite or enhanced satiety – can determine whether a woman loses or gains weight when she exercises. Once again, appetite is the key.

Just like eating, exercise exerts a wide range of effects on the mind. Running, for example, stimulates secretion of brain endocannabinoids which alter appetite, and which may also cause the 'highs' experienced by many runners. In fact, this endocannabinoid response appears to be particularly well developed in animals which evolved to spend a great deal of their time running, such as humans and dogs, and is less developed in other species. Thus humans may be preconfigured to find exercise mood-enhancing. However, despite the fact that women who exercise are often thinner, healthier and more toned, they also tend to have higher rates of eating disorders. This could be because exercising often introduces women to a new cohort of unusually toned and slim women with whom they may subsequently compare themselves, or physical exertion may involve women in activities which place great importance on thinness or prettiness. It is probably for this reason that ballet dancers have lower body-satisfaction than other dancers, and women who take part in 'aesthetic sports' such as figure skating or synchronised swimming are more body-dissatisfied than women in other sports.

Indeed, there is some controversy about how benign the lure of the sporting physique actually is. After the 2012 London Olympics, for example, there was a British media frenzy focused on the physical attractiveness of certain female athletes, with the implication that with their lean, sinewy bodies, they were more healthy role models than the fragile waifs which had preceded them. But many argued that this trend was simply replacing one fatless ideal body type with another, no less unattainable for most women, while further reinforcing the idea that adiposity and curvaceousness are signs of laziness and lack of control.

Of course taking exercise *is* usually a good thing, but it does have its downside too. It can be used to compensate for 'failures of willpower' during a diet – a 'cheat' which, if repeated, can eventually lead to disordered patterns of eating. Also, exercise is time-consuming, and it can have an addictive dimension too – the desire to keep getting that 'exerciser's high'. Although the beguiling nature of exercise may encourage women to persist with their weight-loss regimen, for some women exercise can become the dominant force in their lives. As we will see in the next chapter, compulsive exercise can be a route to psychological withdrawal from the rest of the world, into an activity which is easier than dieting to justify to concerned friends, relatives and partners.

Women did not evolve to live the developed twenty-first century life. Their body shape, size, appetite, self-image and happiness are locked together in a way that worked well twenty thousand years ago, but seems ill-suited to modern life. Women are expected to eat yet not eat, experience food-hedonism and food-guilt, be curvy as well as thin, and to exercise but not exercise too much. Suddenly, having shape, food and mood all bound together inside our heads does not seem like such a good idea.

Yet still the continual messages bombard us that slenderness represents control, and that women's reward for that control is preferential treatment by others. Studies show that slimmer women are more likely to be admitted to university, and more likely to be accepted for a rental tenancy. We even know that women's minds work differently if they think their slimness is being judged – one study showed that women asked to try on swimsuits were subsequently more likely to experience body-shame, tended to eat less, and performed less well at maths tests than women who were asked to try on knitwear.

Women are now dislocated from their shapely, curvy bodies. No longer are those bodies comfortable spaces from which to enjoy the outside world: instead they have been compulsorily commandeered as personal advertisements to that world. Women are trained to adopt others' opinions about how their bodies look and how they *should* look, and they are told that if they want to be happy, then those bodies must be small, trim and without appetite.

SEVEN
A malaise of shapes

Shame is a soul-eating emotion.

The 'Red Book' or Liber Novus, Carl Jung, possibly between 1915 and 1930

'When I went into the fashion industry, I was told I was gigantic, and that my body was letting my face down. I started to resent my body – it was like my head and my body were two separate entities and they were at war with each other. And I met people who were doing insane things to keep looking the way they did, but over time I came to realise that this is what you had to do to be as thin as they are. At first it was an eating disorder I pieced together by copying the people around me. Over time it became a way to punish myself for what was going on in my head. And that's when I really got myself into trouble – it was more like self-harm. At one point I had a hole in the roof of my mouth that had been eroded by vomiting. I was in hospital twice – once because I was so dehydrated and once because there was a tear in my oesophagus.'

Anonymous interviewee 'B'

Eating disorders seem like a step away from the real world: an escape into a way of thinking which is entirely alien to most of us. As a whole they seem completely self-defeating, yet they unnerve us because they carry within them little nuggets of logic – tiny slivers of the behaviours we all use to pick our way through the tangle of food, mood and shape.

Few women eat blindly and without consideration – almost all think about what effects the food they eat will have on their body shape.

Wanting to moderate one's diet and shape to the 'right' levels is almost universal, but when does that desire become pathological? Is there a clear boundary beyond which things get out of control and spiral into uncontrollable starvation or bingeing? Or is there a continuous spectrum of behaviour between 'normal' and 'pathological' eating with no clear demarcation between the two?

Many women assume such a clear boundary exists, and like to think it presents a reassuring barrier to prevent them accidentally slipping across to the other side. Also, a clear distinction between 'normal' and 'disordered' eating would imply that eating disorders are caused by clear-cut genetic, neurological or psychological factors, rather than intangible social and cultural influences which could trip up any woman at any time. Life would be less worrying if eating disorders were neat, distinct abnormalities which can affect only an unlucky few, yet in recent years the suspicion has deepened that eating disorders are, in fact, very messy indeed.

My aim in this chapter is not to provide an exhaustive account of eating disorders, but rather to address a few especially disquieting questions which relate to my wider investigation of female body shape. Are eating disorders entirely different from normal patterns of eating? Do they have a simple biological origin somewhere in the body or brain, or is there more to them than that? Why do they predominantly affect women? And why on earth did the human species evolve to allow such futilely destructive suffering?

It is difficult to formulate a catch-all definition for eating disorders. For a start, they involve a failure to cope with one's own body image. This is something they share with Body Dysmorphic Disorder, but while that disorder singles out particular parts of the body for criticism and distress, eating disorders are directed at the body's global shape and size, and the role of food in creating it.

There are several different eating disorders, but I will concentrate on anorexia nervosa and bulimia nervosa. The words 'anorexia' and 'bulimia' mean 'not eating' and 'ravenous eating', and 'nervosa' simply reflects the assumption that their basic cause lies inside the brain. Anorexia and bulimia are linked, share some features in common, and as many as a third of sufferers may migrate from one disorder to the other, yet they also exhibit differences which set them apart from each other.

Many psychiatrists take a rather 'tick-box' approach to diagnosing illness, and their four defining characteristics required for a diagnosis of anorexia are extremely low weight, a cessation of menstrual periods, a fear of gaining weight, and a distortion of perceptions of body weight. Anorexia sufferers may reach and maintain their low weight by restricting their eating, or by purging any food they do eat by vomiting or taking laxatives. They often show an obsession with a restricted set of particular foods, may hoard food, eat extremely slowly, may not swallow food, or may regurgitate it to rechew it or discard it. They often take steps to conceal their behaviour, and may wear baggy clothes or adopt particular postures to hide their body shape. They are prone to weakness and dizziness, and suffer from hormonal disturbances which can lead to, among other things, stunted growth and osteoporosis. Anorexia nervosa probably has the highest suicide rate of any mental illness.

Bulimia nervosa is rather different. Its 'tick-box' definition has three elements – bingeing with high-calorie foods at least twice a week; 'cancelling out' those calories by purging, laxatives, enemas, starvation or exercise; and a strong tendency for self-worth to be based on body size or shape. On average, bulimia starts at greater age than anorexia, and sufferers are usually a fairly normal weight – they fear gaining weight, but do not particularly strive to lose it. The binges may contain between seven and ten thousand calories, but this varies a great deal – what seems to be important is that they are *perceived* as binges. Most bulimia sufferers induce vomiting to purge food, often more frequently than once a day, and this leads to most of the adverse physical effects of the disease – calluses form on the knuckles, gastric acid demineralises tooth enamel, electrolyte imbalances disturb the normal heart rhythm, and the stomach and oesophagus may rupture.

At first sight, the behaviours associated with eating disorders may appear completely alien to other people, yet the distinction between normal and abnormal is not at all clear. First of all, there exists another diagnostic category called 'eating disorder not otherwise specified' which includes people who exhibit abnormal eating behaviours but do not meet all the criteria for a formal diagnosis of anorexia or bulimia. While this probably just represents the limitations of using a crude 'tick-box' diagnostic system, more worrying is the idea of 'partial eating disorders' – an undefined but potentially large group of people who display some of the symptoms of anorexia or bulimia, sometimes for many years. I suspect that it is this hidden reservoir of uncounted sufferers which makes the published figures for the incidence of eating disorders seem strangely low – 0.1–0.5 per cent of the population with anorexia, and 0.5–1.5 per cent with bulimia. Certainly, when interviewing women for this book, every single one of them knew a close female friend or a relative with an eating disorder.

One further feature of eating disorders deserves particular attention because, as we will see, it may relate to the possible origins of these diseases: exercise. Many people diagnosed with eating disorders, and many women who are simply dissatisfied with their bodies, take a great deal of vigorous exercise in an attempt to change their shape. And their relationship with exercise changes fundamentally: 'exercise dependence' is diagnosed when the main aim of exercise is to induce weight loss or shape change, and when an exerciser feels guilty when they do not exercise. Many people with exercise dependence, especially women, will exercise while injured, or miss out on social events because they are exercising. Some women become exercise dependent as part of an eating disorder, but for many it works the other way round – they develop an eating disorder as a result of their exercise dependence.

Although different studies vary in their estimates of how common eating disorders are, they all agree on one thing: they are approximately ten times more common in women than men. The fact that some men do get these disorders is helpful in our attempts to understand them, but the existence of this sex-skew in a species with such an unusual female body shape is suspicious, to say the least.

Just as the symptoms of eating disorders are extremely variable, so is the course of these diseases. Some women suffer a single, brief episode, whereas others live with an eating disorder for decades. A 'typical' eating disorder may last for six years or so. Many, but certainly not all eating disorders start during the second decade of life, and their incidence slowly decreases thereafter. Eating disorders tend to start earlier in girls who undergo puberty earlier, but by the age of twenty those girls are no more likely to have an eating disorder than their later-developing counterparts. Eating disorders dramatically reduce fertility, but if sufferers do become pregnant it appears that, on average, the symptoms of eating disorders recede, and may even disappear. Wherever eating disorders are concerned, however, there are always exceptions – and some women suffer from an eating disorder, 'pregorexia', in which they starve or purge in an attempt to prevent the (natural) weight gain that comes with pregnancy. After pregnancy, some women report that their eating disorder remains quiescent while they are caring for their children, but reverts to its original severity once those children leave home. Eating disorders are less common in middle and old age, but they are not rare. Sufferers in these age groups have been largely ignored by the media, who presumably find diseases of older women less tragic, and by researchers, who often base their research on a convenient supply of female university students.

So eating disorders are strange, dangerous, intriguingly variable, either rare or common depending on how you define them, and predominantly affect women. They also involve patterns of disordered thought which make them look like mental illnesses. This, and the desperate desire for successful medical treatments, have led to an enormous amount of recent research into the biological basis of eating disorders. In the past, scientific studies have shown us that many other diseases have single biological, organic causes, and this knowledge has often led to miraculous cures, but things have not turned out that way for eating disorders. Not yet, at least.

There is good evidence that eating disorders have a heritable genetic basis. Studies of identical and non-identical twins who are raised together or apart, and statistical analyses within families, suggest that genes could be the biggest single factor underlying eating disorders. An individual is much more likely to suffer from an eating disorder if they have a close relative who is also a sufferer, and this effect is not explained by them living in a similar environment. Unaffected family members tend to share many of the character traits which predispose to anorexia or bulimia. Even traits as apparently complex as the tendency to relate one's shape and weight concerns to one's self-worth have been shown to be strongly inherited. In striking contrast, childhood eating disorders (which are rarer) do not appear to have a heritable basis at all.

Although finding the individual genes behind inherited eating disorders has proved frustrating, the general power of genes seems very clear and those that have been identified so far are a thought-provoking bunch. The best candidate gene is 'SLC6A4' which is involved in the transport of serotonin – the brain chemical we saw in the last chapter is involved in mood and appetite. However, although the statistical link between variants of this gene and eating disorders is clear, it is not powerful – variations in the gene exert only small effects on women's chances of developing eating disorders. Another candidate gene, 'COMT', is involved in the breakdown of brain chemicals similar to serotonin, such as dopamine, while the 'ESR' genes, which produce the receptor proteins which allow oestrogens to act on cells, have also been implicated. Despite this initial success, we should not expect this gene-search to produce simple answers – after all, any single, simple gene defect which caused something as damaging as an eating disorder would have been lost from the human population long ago.

As well as genes, personality traits have also received a great deal of attention as possible causes of eating disorders. Girls and women who have eating disorders tend to be perfectionists, to be anxious, and to display psychological rigidity in the face of change – all characteristics which can be measured by objective tests. It has also been claimed that they are more likely to show characteristics akin to autism. This does not mean that eating disorders and autism are the same thing, but it is notable that people with both conditions often see their 'condition' as 'part of themselves', rather than as a disorder or disability.

Studying the role of personality in eating disorders has led to two important realisations. The first is that these personality traits are demonstrably present *before*, and often long before any behavioural, psychological or physical symptoms occur. The fact that they pre-date any starvation or bingeing means that, although these traits may be exacerbated by the severe physical effects of eating disorders, they are not caused by them. The second realisation is that 'control' does not seem to feature highly in scientific analyses of eating disorders. We are often told that anorexic girls behave as they do because diet is the one thing in their lives they feel they can control, yet the evidence to support this is lacking. Indeed, I have spoken to anorexia sufferers who were adamant that their condition involved an unpredictable, unplanned and shockingly sudden *loss* of control over food. Admittedly, many pro-anorexia and pro-bulimia websites strongly promote the idea of control to their readers, but 'control' is not the universal initiating factor it is often claimed to be.

Some of the most intriguing insights into eating disorders have come from studies of brain function. These studies provide very real and specific insights into the mind of anorexic or bulimic women, which go some way to explaining the unusual nature of these disorders, and they also raise the possibility of developing targeted medical treatments.

However, before I describe them, I should mention one important caveat regarding these neurological findings. Because definitively diagnosed eating disorders are quite rare, it is extremely difficult to conduct a study in which detailed neurological data are gathered from women *before* they develop these conditions. Even a large study which follows hundreds of girls through childhood and puberty is likely to end up with fewer than ten girls with tick-box-diagnosed eating disorders – probably not enough to provide any meaningful information. Because of this, all the brain alterations so far associated with eating disorders could be the *result* of starvation or bingeing or purging, rather than the cause. For example, the brains of women with anorexia are smaller and contain larger internal cavities, but this is probably because anorexia causes the brain to shrink, rather than the other way around.

The first cognitive abnormality seen in eating disorders involves abnormal *perceptions* of body shape and size, and indeed in some cases this could be the key to the entire disorder. Sometimes women with eating disorders recount moving experiences in which they accidentally glimpsed a woman whom they considered to be abnormally and unpleasantly thin, only to discover that they were unknowingly looking in a mirror. Such body shape 'epiphanies' are evidence that some women with eating disorders drastically overestimate their own body size due to an inherent abnormality of self-perception. And experiments with computer-distorted images of sufferers' own bodies confirm that, *on average*, anorexic and bulimic women do indeed overestimate the real size of their own bodies. However, size overestimation is not a consistent finding in all women with eating disorders, so it is not the defining feature of these conditions that it is sometimes suggested to be.

Such perceptual deficits appear to be specific to bodies and food. Women with distortions of body perception are usually able to accurately assess the size of neutral objects, such as vases. For example, brain scans show that the cerebral cortex in anorexic women is less activated by images of food or other women's bodies, suggesting that they are selectively 'blunted' to these stimuli. Conversely, when women with anorexia view images in which their own body shape is computer-distorted to look larger, one brain region, the dorsolateral prefrontal cortex (see picture) becomes unusually active – and the more severe their anorexia symptoms, the more excessive this overreaction is. Brain regions known as the insula, hypothalamus and amygdala also exhibit reduced responses when women with anorexia are asked to think about eating, while the insula also responds more weakly to the taste of high-calorie foods in the mouth. The insula is probably involved in seeking out desirable foods, but as we have seen previously, it may also play a role in women's sense of body-ownership.

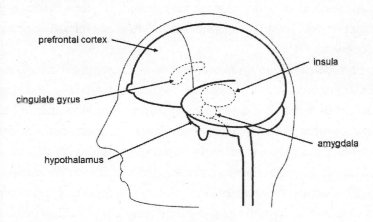

Some perceptual abnormalities in eating disorders appear more complex. Studies in which women's gazes were tracked as they looked at their own bodies and the bodies of others show that eating disorders entail an unusual selectivity about what is being looked at. Women without eating disorders spend more time looking at the parts of their own bodies which they perceive as 'beautiful' than parts they find 'ugly', and they do the opposite when looking at other women's bodies. In contrast, women with eating disorders concentrate on 'ugly' bits of themselves and 'beautiful' bits of other women. And perception in eating disorders is also mysteriously linked to apparently unrelated aspects of brain function – for example, women who are strongly biased towards being right-handed are more likely to perceive their own body shape inaccurately and, as a result, exhibit symptoms of disordered eating.

The second group of brain abnormalities involves *cognition* – higher level processing of information about the body, including interpreting and assigning importance to its shape and size. For example, some anorexic women perceive their body size correctly, yet still believe it is not too thin. Also, many women with eating disorders have overly strict ideas of what constitutes an 'ideal' body, or over-emphasise the importance of body shape to their self-worth, happiness and success. A quick internet search for 'pro-ana' websites quickly demonstrates how some anorexic girls and women believe that bones are white and clean and beautiful and should be seen through your skin, whereas fat is disgusting and evil and makes everyone, including boys, hate you.

It may come as no surprise that women with eating disorders exhibit skewed cognitive responses to bodies and food, but eating disorders also seem to involve more general abnormalities of high-level thought. For example, women with anorexia perform less well on tests of almost all aspects of higher cognitive functioning, including memory, attention, abstraction, visuo-spatial ability and synthesising information from different sources. Also, just thinking about eating food induces abnormally exaggerated responses in the prefrontal cortex of women with anorexia, almost suggesting that they 'over-think' food with their higher cognitive areas, rather than savouring it with their baser brain regions. Many of these cognitive abnormalities persist even after eating disorder episodes have ended, which raises the possibility that they may also have been present before the disorders ever started.

The third and last set of brain abnormalities seen in eating disorders involve altered *emotions* – especially shame, disgust, anxiety and self-consciousness caused by food or the idea of weight gain. Many of us experience mild versions of these emotions quite frequently, but they come to dominate the lives of women with eating disorders. Food and weight can also independently affect our emotions in different ways – many children, for example, are repelled by the idea of eating certain foods, but not usually because they think they will make them fat.

In recent years, neuroscientists have actually located the regions of the brain involved in emotion – a list of regions which partially overlaps with areas we have encountered already: the insula, amygdala, prefrontal cortex and cingulate gyrus. It had been suspected for some time that women with eating disorders exhibit more negative emotions when standing in front of a mirror, or have trouble interpreting social cues from the people around them, but we can now watch these responses taking place in real time, inside living brains. For example, the amygdala, a region involved in fear and disgust, seems to be more active in women with anorexia. Also, comparison of pictures of oneself and others' bodies leads to distorted patterns of activation throughout the 'emotional' parts of the brain.

Amine chemicals in the brain, such as serotonin and dopamine are increasingly implicated in emotional abnormalities in eating disorders. These chemicals underlie diverse brain networks involved in, among other things, mood, appetite and impulsivity – and their patterns of activity are obviously altered in both anorexia and bulimia. The effects of these chemicals are extremely complicated, which probably explains why crude attempts to use serotonin-altering antidepressants to treat eating disorders have proved rather unsuccessful. Instead, alterations in these chemical systems should perhaps be seen as reflecting general shifts in the complex way the mind thinks about food – a tussle between the ancient brain centres of hunger, appetite and emotion, and the more recently evolved regions of 'higher' thought and control.

So the biology of eating disorders remains confusing. They have a clear genetic basis but we cannot find many good candidate genes; there are predisposing character traits which underlie them but we do not know why; and despite many striking neurological findings we have discovered no single brain abnormality which *causes* anorexia or bulimia. This lack of success could be because we simply have not yet looked hard enough, or maybe, all along, biology was the wrong place to look.

And some claim that biology is indeed the wrong place. Instead of a gene, or a brain abnormality, should we instead be seeking social or cultural explanations for eating disorders? Environmental triggers and cultural pressures may sound wishy-washy, but they can be powerful.

Some women recollect a specific environmental cue which they believe drove them to initiate their disordered eating and calorie purging. Conversely, many women can recall no such cue – their anorexia or bulimia seemed to come entirely out of the blue. And even for women who do believe there was a specific trigger for their disorder, the variety of possible cues is bewildering. Often it is thought that exposure to a single image of a female body – either their own or someone else's – set them on a one-way road to abnormal eating habits. Others believe that being overweight as a child was the trigger (and there is some evidence that this can indeed be the case). Perhaps surprisingly, factors such as parental death or childhood sexual abuse do not seem to correlate with subsequent incidence of eating disorders.

Being teased about one's body shape has often been linked to the onset of eating disorders, although such teasing is so common that this is hard to prove – one study suggested that 13 per cent of girls are teased about their body shape by their mothers, 19 per cent by their fathers and 29 per cent by their siblings. Some girls even thought that their mother's eating habits were a contributing factor. And most worryingly of all, many women believe it was an initial, apparently innocuous, teenage attempt to diet which got them hooked on controlling their food intake and body shape.

There is a persuasive argument that, rather than providing single provoking triggers for eating disorders, modern society and culture impose a global obsession with feminine thinness and food-denial which both causes and perpetuates eating disorders. In other words, women with anorexia or bulimia are simply individuals who happen to lie at the vulnerable end of the spectrum of women's reactions to the cult of thinness in which they all find themselves. So perhaps eating disorders are simply an exaggerated form of the shape-sickness which afflicts society.

With that in mind, there may be several reasons why girls and young women develop pathological eating and shape-control habits, but all of them involve attempts to challenge external expectations of the female body. Girls often start to show signs of eating disorders around the time of puberty – a phase when they are being pulled away from the relatively androgynous slim body shape of childhood and towards a more curvaceous, womanly figure. Alternatively, eating disorders could be a way to reject the feminising effects of puberty – a rejection of curves – or an attempt to recreate the blissful sexlessness of infancy when girls and boys did not seem very different. Maybe in a culture in which women are often judged solely by their distinctive body shape, girls believe that by actively expunging that shape they can liberate themselves so they can be appreciated for their achievements instead. Or possibly, when society frowns on teenage sex for girls (but not boys), could eating disorders be a way to avoid sex?

The idea that our modern developed society and culture cause eating disorders is disturbing, partly because it forces us to face the extreme pressures we put on women, especially young women, and partly because it makes us realise that many of the 'distorted' attitudes of women with eating disorders are, in fact, valid – people *do* treat slimmer women better, they *do* think more of them, and women *can* become slimmer by manipulating calories. Accepting a social cause for eating disorders also implies that no one is safe from these malign influences: you can evade a bad gene, but you cannot evade society. And women do not have to be diagnosed with a disorder to feel these pressures – many women who consider themselves 'normal' engage in some fairly drastic methods of weight control.

Supporters of the social theory claim that it explains many strange features of eating disorders – especially why they are more common in women living in developed countries, and young, intelligent, educated, affluent women at that, and why they have become increasingly common in recent decades as the cult of thinness has taken hold. Indeed, these are often seen as the defining social features of anorexia and bulimia, yet whether they do indeed preferentially affect the rich, developed and educated is still open to debate.

Similarly, although some medical data might make us believe that anorexia and bulimia are essentially modern disorders which only became common within the last fifty years, this may reflect changes in attitudes or diagnostic criteria rather than actual prevalence of the disorders. Anorexia was first medically described in 1868 (as 'apepsia hysterica') and bulimia, remarkably, not until 1979. However, closer inspection of written records has led many to claim that both conditions can be traced back to the Middle Ages, and even before, possibly with rates of incidence just as high as today. 'Holy fasting' was commonly reported in girls in medieval Europe, and many died as a result. Indeed, many Christian saints were young women who starved themselves to death in the name of God.

Thus, while social and cultural causes could explain socio-economic and historical variations in anorexia and bulimia, first we do need to be clear that those variations actually exist. Also, at first sight, these theories cannot explain why men sometimes get these purportedly culture-inflicted, female-focused disorders. Unlike girls, the bodily restructuring of puberty is almost universally welcomed by boys – boys generally want to be tall, muscular and deep-voiced – so what cultural pressures are *they* reacting against? Some have pointed out that many male eating disorder sufferers are homosexual or make their living from dancing or acting, and that they thus represent unusual male subcultures to which different rules apply. However, there is evidence that eating disorders are becoming more common in the male population as a whole too, and some have ascribed this increase to a greater prevalence of images of fatless male bodies in the media – exactly as is claimed for eating disorders in women.

The most confusing aspect of eating disorders is why they occur at all – how humans evolved into creatures who could suffer them. This is the side of anorexia and bulimia which interests me most, because it strikes at the heart of what it means to be a human female and to balance the ancient conflicting demands of food, shape and success in a modern, unnatural world.

No animal evolves specifically to suffer bouts of starvation, bingeing, emaciation, infertility and death, yet that is precisely what some women now do. However, all current human behaviours occur in the context of our evolutionary inheritance from the last few million years. Even if full-blown eating disorders have only existed for the last five decades (which I suspect is unlikely), then they must still be the product of a female brain honed by aeons of natural selection. The socio-cultural pressures on women today must act on *something* – and that something is the physical and mental adaptations which once helped their prehistoric female ancestors survive and thrive. This does not mean that eating disorders are advantageous today, but I strongly believe that *elements* of them must have proved advantageous at some point in the 99 per cent of our history when we were hunter-gatherers. Anorexia and bulimia did not appear out of nowhere.

There are two main types of evolutionary theory which attempt to explain the origins of eating disorders. The first set of theories emphasise the idea that food supply was intermittent for much of human history – that our ancestors experienced alternating periods of plenty and scarcity. This could have been because some foods were only seasonally available, or it could just be because climate and food supply were generally unpredictable.

One theory posits that alternating episodes of bingeing and starvation were a normal feature of pre-agricultural human life, and that we retain a propensity for this behaviour today. This could certainly explain why bingeing is a natural behaviour – storing valuable calories when times are good – but self-starvation seems harder to explain. However, there is clear evidence from the natural world that abstinence from food can be a normal, inbuilt urge too. For example, many animals reduce their appetite *in anticipation* of food scarcity – well-fed deer eat less in the winter simply because their ancestors lived in an environment where food was scarce at that time of year. In other words, animals do not waste energy seeking food if their genes programme them that it is not to be found. So it is entirely possible that both the bingeing and starvation seen in eating disorders could be evolutionary relics of a time when our food supply was unpredictable.

Another, perhaps opposing theory based on erratic food supply relates to the over-exercising and hyperactivity seen in some eating disorders – women with anorexia, for example, often pace around unnecessarily to burn off extra calories. According to this theory, hyperactivity is a normal response to food scarcity which encourages famine-stricken human populations to move, literally, to pastures new. And seeing a few emaciated women in their midst may have been enough to convince ancient tribes that it was time to move or die. Thus women with eating disorders were like canaries in a coal mine – hypersensitive individuals who warned of impending doom. There are some who suggest that anorexic women were actually the people who energetically uprooted their starving yet lethargic kinsmen and led them to the promised land. This is a strange notion, and it obviously does not explain why so many women *self*-starve when food is plentiful. However, there is experimental evidence that rodents, too, become overactive and increase food-seeking behaviour when starved, although well-fed lab rats do not spontaneously become anorexic in the first place.

The final food-supply-based theory puts even more emphasis on the importance of human social groups. Indeed, it is predicated on the assumption that being a member of a tribe or social group is absolutely essential for a woman's survival. This hypothesis suggests that for some women, eating less is a natural response to reduced food availability, which stops them competing with other people in their social group and thus risking expulsion. This idea seems strange at first sight, but it must be admitted that we do not know what awful things ancient human tribes might have done to individual members in their desperate attempts to avoid starvation. This theory has the added advantage that it could explain why women *start* to restrict their eating in the first place. It also suggests that women may *continue* to under-eat even once food supply has increased, to avoid socially threatening competition for food with men who are larger, angrier, and presumably now less starvation-weakened than them. According to this idea, exclusion from the tribe is a greater risk than under-eating.

The second group of evolutionary theories for eating disorders relates to reproduction. Certainly all animals must have pre-programmed instincts to stop eating – otherwise they would never have time to breed, or perform other essential activities. This of course could explain why eating disorders often start during puberty, as this is a time when teenagers' thoughts naturally turn to sex. However, paradoxically, eating disorders usually lead to reduced sexual activity, so could they be freeing up girls' time to do other things instead? One possibility is that they represent an unconscious strategy, usually successful, to delay puberty or halt menstruation. In the past, this may have been a sensible mechanism to avoid a potentially disastrous pregnancy during times of want – although self-starvation seems an unnecessarily risky way of achieving this. And today, girls face tremendous pressures not to get pregnant, to succeed in school, university and career, so it is certainly possible that these modern stresses now trigger those ancient female strategies for reproductive restraint.

Another reproduction-oriented theory relates to competition between women for male mates. Although this may not seem a very feminist idea, it is claimed that in societies in which women compete for male attention by appearing slim and youthful, some individuals may 'over-compete' by becoming extremely thin and losing the curves which come with maturity. Thus eating disorders could represent an abnormally exaggerated psychological response to prevailing body-shape ideals – some women get slim, but others go too far. And there is experimental evidence that women do indeed eat less after meeting women with high social status, even if those women are not themselves particularly slim.

The final evolutionary theory of eating disorders is perhaps the weirdest of all, because it has a bizarrely self-sacrificing flavour to it. The idea is that girls in high-status, mutually protective, families stop eating to free up resources for their siblings – and their brothers in particular. Because high-status boys have the potential to father more children than their sisters could mother, it is argued that sometimes the best way for a girl to pass on her genes is to support her brother in his attempts to sow his seed far and wide. After all, the arithmetic of inheritance dictates that siblings share approximately half their genes, so if a girl's brother sires three children (which can be the work of just a few hours for him), then more of her genes will reach the next generation than if she rears a single child herself (which is the work of two decades for her). It may seem utterly perverse that girls should self-starve to add fuel to their brother's promiscuity, but the genetic numbers would, in fact, add up in support of this theory.

All these evolutionary explanations of eating disorders sound, to be honest, weird. We are startled by the suggestion that women contain within them a genetic propensity to self-starve so they can cope with an erratic food supply, signal impending starvation to others, evade social conflict, avoid pregnancy, attract a mate, or encourage their brothers to sleep around. All these ideas have their weaknesses, and no single theory can explain every aspect of eating disorders. However, some of them probably contain shreds of truth, and none is mutually exclusive, so maybe those shreds have combined to underpin the eating disorders which blight so many women's lives today. After all, there must be *something* which makes females of our species, and our species alone, prone to these debilitating conditions.

I would speculate that the sheer harshness of life during our evolutionary history has indeed left women with a complex set of inbuilt responses to famine and plenty. And those responses were helpful when we were hunter-gatherers, but do not fit well with modern life where food is not just abundant but over-abundant. Eating the right amount at the right time was truly a matter of life and death for women over much of the last few million years – consuming enough to allow them to carry out their roles in prehistoric societies, while deferring to the energy needs of calorie-guzzling men. So women evolved robust, almost stubborn strategies to cope with these horrendous pressures, and that stubbornness is evident today in the immense resistance to treatment of many eating disorders.

Those stubborn strategies are built into the vast human female brain as a complex network of neural reactions to food, mood and body shape, and like so many things in biology, those neural networks *differ* between women. There is no single best strategy, and different women cope in different ways. I believe that the neurologists are correct, and that some women possess quirks of brain circuitry which predispose them to eating disorders. But I also think the sociologists are correct too, and that the precipitating cause of anorexia and bulimia is usually the culturally imposed idea that thinness is good. And in the next chapter I will examine whether the relentless barrage of skinny images, thoughtless comments and makeover shows really does affect how women think about body shape.

If a woman has a genetic or neural predisposition towards eating disorders, then all it may take is a twinge of worry about her body shape and an apparently harmless, 'well-intentioned' diet to precipitate her slide into anorexia or bulimia. I would argue that women's hyper-complex brains provide the vulnerability, and the cult of thinness provides the trigger.

Female body shape has a central importance in the workings of the female mind, and the interactions between the two are fiendishly complex and deeply ingrained. They explain so much about what it means to be a woman, living in a woman's body, with a woman's appetites, yet the body and the mind do not coexist in isolation. As we will see in the last part of this book, they cannot ignore the world outside.

PART III

THE WORLD

'My best friend is one of the leaders of my group. She's intelligent, good
looking, she looks like a model, her legs are six feet long – her legs are
nearly as tall as me. She is confident around women and men. Another
of my friends is more of a follower – she is not at all body confident, she's
overweight, she's unsure of herself, so I guess that's why she looks up to my
best friend – why she's taken a submissive role.'

Anonymous interviewee 'A' (age 21, body mass index 22.7)

'I think it's because my mum is very funky and glamorous and beautiful and
slim and elegant. People would just define her as being gorgeous and I didn't
want to be that person, so I actively rebelled against it. Also, no one told me I
was beautiful as a kid. I had no relationship with my body whatsoever.'

Anonymous interviewee 'B' (age 32, body mass index 26.1)

'I don't feel content living in my body. I hate it. However much I work
on it – it's not what I want it to be. I can't think of a positive thing to
say about it. And I don't think anyone can say anything which would
make me happy with it.'

Anonymous interviewee 'C' (age 33, body mass index 21.3)

'It was the worst industry I could go into – interviewing all those pop stars, all glamorous and perfect. But I know it's all smoke and mirrors – I've been on the photoshoots with these people. Everyone knows that (——) looks flawless and amazing in pictures, but when they take off the body makeup she cries. Yet there's still a part of me that buys into it. I know the reality of it, but I still like to torture myself with it.'

Anonymous interviewee 'D' (age 40, body mass index 24.5)

'I enjoy getting dressed up, I enjoy the creation of the person who's going out. I'm quite pleased – it sounds awfully immodest – but I like to look like something I enjoy the look of. Now I would say that I have a better body than most of my friends – but how does that make me feel? It makes me feel quite pleased – but that makes me sound competitive.'

Anonymous interviewee 'E' (age 70, body mass index 24.8)

EIGHT
Following the fashion

Meet Feral Cheryl! Here she is, the anti-Barbie, fresh from the rainforests of Australia. This 34 cm vinyl doll runs barefoot, dreadlocks her hair with coloured braids and beads, wears simple rainbow clothes, has piercings and a range of tattoos, and even a bit of natural body hair.

www.feralcheryl.com.au

'I think most women look in the mirror for reassurance that they look nice. I tend to look in the mirror for confirmation that I look as bad as I think I look.'

Anonymous interviewee 'C'

'I'm a Barbie girl, in a Barbie world' explained Lene Nystrøm, the lead singer of Danish-Norwegian pop group Aqua, in their 1997 UK number-one single. She spoke of a 'life in plastic', considered it 'fantastic', exhorting the listener to brush her hair, and undress her everywhere. It was also noted that she was versatile enough to simultaneously 'act like a star' and 'beg on [her] knees'. Despite the enthusiastic, breezy nature of the song, the creators of the Barbie doll, Mattel, filed a lawsuit against Aqua because they believed that elements of the lyrics did not chime with the ethos of their product. Eventually the acrimonious suit was dismissed, and the presiding judge suggested, 'The parties are advised to chill.'

Although the shape of Barbie has varied since she was launched in 1959, her waist–hip ratio has usually been approximately 0.59 – an exaggerated version of the feminine shape, rarely seen in nature. In contrast, according to my detailed measurements of images of Lene Nystrøm, she possesses a ratio of approximately 0.75. While this brings her extremely close to the value considered optimal by both men and women, it suggests that she is not, as she claimed, a Barbie girl.

At the time of writing, it is thought that one billion Barbie dolls have been sold, and 'Barbie Girl' is the thirteenth biggest-selling single in UK chart history.

Despite Barbie's grip on us, there is no authoritative definition of society's ideal female body shape. Instead, it is an agglomeration of the ideas we all hold in our heads and in the printed and electronic images our society so reveres. Although it can be hard to pin down, we each have a fair idea of what the current ideal body shape is, and that ideal is extremely powerful. Many women want to look like it, or at least they want to look *more* like it, and many heterosexual men assume they should consider it when deciding whom they desire.

At present, the Western societal ideal for a woman's shape is slim – slimmer than most women are, certainly. Something in our environment keeps telling us that slimness is healthy, normal, attractive and achievable if women are prepared to put in a bit of effort. Of course there is a societal ideal for men's shapes too, but there does not seem to be as much pressure on them to attain it. This could be because there are fewer pictures of men around us, or maybe men are partially psychologically exempt from this pressure – for reasons to which I will return in the last chapter. Certainly, the extent to which a woman matches the female ideal is often assumed to reflect her social and economic status, her psychological self-control, and her moral goodness.

Yet where does this 'ideal shape' come from, and what imposes it on us? As we will see, it changes over the decades, and it varies around the world, so how can something so fickle, so parochial, so here-today-gone-tomorrow, be so very important?

It is often claimed that the prehistoric norm was for ideal female shapes to be curvaceous, even obese. Indeed, there are several examples of prehistoric art from around the world in which women are depicted in just such a way – with huge breasts, bellies, buttocks and thighs. A famous example is the 25,000-year-old 'Venus of Willendorf' statuette from Austria (*see figure below*), but similarly curvaceous artefacts have been discovered elsewhere.

However, these depictions do not necessarily represent an archetypal ideal for the female form. There is no evidence that women this shape existed at the time they were created, and it is possible that the shapes were exaggerated for some particular reason – to make it clear that they were indeed female, or possibly as a depiction of pregnancy. After all, there are certainly ancient statues of female bodies in which the aesthetic aim is exaggeration rather than accuracy – the statue of Artemis from Ephesus has eighteen breasts and this probably signified something important, but I doubt it shows that Ephesian men desired multi-breasted women. Calling any ancient curvaceous depiction a 'Venus' also involves a big assumption about

what those curves were for. They might have been a man's ideal, but they could just as easily have been a woman's self-portrait.

Ancient figurines of slim women also exist. For example, the Cycladic civilisation of the Aegean was doggedly churning out huge numbers of flat, almost geometric representations of slim female bodies between five and four thousand years ago (*see figure above*). These figurines are obviously female, because they possess a vulva and small, conical breasts. They also usually have their forearms crossed across their bellies, perhaps to cover a pregnant bump (which is visible in some of them), or as a representation of period pains, perhaps. There is often, but not always, a narrowing at the waist, which gives the thighs a slight curve, but that is as curvy as these figures get. Some even have an 'inverted triangular' torso with broad shoulders, not unlike the current Western male ideal body shape. No one has ever called these figurines 'Venuses', and yet they were far more abundant and ubiquitous than most of the curvy statues.

From the Western classical period to the start of the nineteenth century, women represented in art were, almost without exception, quite pear-shaped in comparison to today's actresses and models. They usually had curvaceous buttocks, thighs and a clear indication that their bellies were curvy too. Indeed, sometimes the artistic urge to make the female abdomen appear fertile made it difficult to tell whether women in Western art were meant to be pregnant or not. Eve, especially, as the mother of humanity was often permitted an equivocal bump. Breasts in art were, at this stage, relatively small – large enough to depict maturity and femaleness, but no larger than that. An exception to all of the above trends was the Virgin Mary because she, by her very nature, was conflicted. She was an ideal woman, obviously, but not *that* sort of ideal woman. She was sometimes quite pear-shaped, and the occasional breast was visible explicitly only for serious, non-sexual Lord-suckling purposes. In other images, she was depicted earlier on in her story, at the Annunciation, as a virginal waif-like character. Then again, everyone looked gaunt and frail in Byzantine icons.

Sometimes, post-renaissance artists would focus on curvier examples of the female form – the early seventeenth-century Flemish painter Rubens being the most famous example – but rarely were women depicted as being slim, unless as an indication of poverty. Indeed, even in the mid-nineteenth century some of Edouard Manet's paintings of women were considered to be unerotic or even obscene because they were considered too thin, although the fact that they blurred the boundaries between holy and lascivious, and were obvious depictions of young contemporary women, probably did not help.

Things became extremely complicated in the twentieth century. We are often told that the Western ideal of female shape has become progressively thinner over the last century or so, but things are not as simple as that. Other researchers have suggested that the ideal started slim, became curvier mid-century, and has slimmed down again in recent decades, but even that is perhaps an over-simplification.

In the 1920s economic times were good in much of the Western world, social restrictions on women were decreasing and in some countries, they were even allowed to vote. This period saw the demise of the physical constraints of the corset, and their replacement by the dietary constraints of a newly slim, almost boyish ideal. And, after centuries of concealment, women's legs also re-emerged, at least to below the knee. Women may have consumed a similar number of calories to today, but they expended more of them in physical activity, so they were slighter – one estimate puts the UK and US average 'vital statistics' for young women at 31-20-32. Women were also smaller in stature, due to childhood diseases and a less copious supply of calcium-rich foods. Despite this, many women still resorted to flattening their breasts with bindings to attain the prevailing ideal.

Between the 1930s and 1950s, from the depression onwards, curvaceousness slowly reappeared, aided by improvements in health and nutrition – and this trend even continued during the Second World War in countries which were not invaded. It has also been suggested that women opted for a more curvy look as sex and relationships changed in response to men's increasingly protracted and perilous absences during the war. Eventually, Marilyn Monroe and Jayne Mansfield came to epitomise this trend. Some even saw this change as a business opportunity, specifically marketing products to women who wanted to put on some pounds to reach their ideal curvy weight. 'Men wouldn't look at me when I was skinny. But since I gained ten pounds this new, easy way, I have all the dates I want', cajoled one advert by the Ironized Yeast Company of Atlanta, Georgia.

Then, in the 1960s, something changed. Along with an economic boom, the contraceptive pill and a claimed improvement in women's social status came a slimming-down of the ideal, exemplified by Audrey Hepburn, Jacqueline Kennedy and Twiggy, and prefigured by Alfred Hitchcock's blonde leading ladies. And the slim ideal has persisted to the present day, but with variations – a more assertive, muscular shape appeared in the 1980s, for example, along with the male-mimicking shoulder pads so characteristic of that era. In the 1990s, although many top actresses and models were not as slight as in the sixties, there was a brief fashion for extremely thin, frail and even ill-looking models, sometimes called 'heroin chic'. In spite of these variations, the body mass index of *Playboy* centrefold models did not change between 1980 and 2000, yet by that year, the average Western vital statistics had increased to perhaps 36-28-38.

Trends like these are often best viewed in retrospect, but I suspect that the early years of the twenty-first century may be seen as a reversal of the trend for thinness, with many famous actresses and singers being noticeably and self-confidently curvaceous. It seems likely that fluctuations over time are the result of a combination of economic and social factors, as well as a legacy of the individual women who happen to catch the public's eye. Certainly, the message from the last hundred years is that the societal ideal of female shape is anything but fixed.

Variation in female body ideals is also evident in different ethnic and social groups – and this is as true of ethnic groups living within the same country as it is of different groups around the world.

Most comparative research has investigated differences in body-shape ideals between black and white ethnic groups in the United States and, to a lesser extent, Europe. Although African-American women are more prone to obesity than white Americans, they have higher levels of body-satisfaction. This is true at any size – underweight, slim, average, heavier, overweight or obese African-American women are all more content with their shape than white women in the same weight-ranges. Intriguingly, their actual perception of their body differs, and they are more likely to think of themselves as smaller than they really are. And of course, this could partly explain why eating disorders are rarer in African-Americans too.

It has been suggested that these differences may partly be explained by African-American girls being more influenced by the opinions of their close family than by the media. Certainly, it is the case that most thin-ideals presented in the media are white – in fact, there are very few positive images of black girls on Western television – so perhaps the slightly curvier black ideal is a benign effect of this media neglect. Indeed, one study has shown that the curvy ideal persists even though black girls are no less likely than white girls to endure childhood teasing for being perceived to be overweight – so there must be *something* which is making them more resilient. Some black women have suggested that for them, grooming, style and confidence are more important than body shape and size, although research so far has not backed up this suggestion.

Another possibility is that black women may be responding to the preferences of black men. Studies do suggest that black men prefer women with slightly lower waist–hip ratios – narrower waists and larger buttocks – than white men, but the data regarding men's weight preferences are conflicting. Experiments in which men were asked to rate images of women's bodies did not suggest that black men have a preference for women with higher body mass indices, nor that they showed greater 'flexibility' – a wider range between the lightest and the heaviest women they would like to date. Yet another study suggests that black men are more likely to express a preference for heavier women in dating ads, and that they are less likely than white men to secretly want their current partner to lose weight.

Thus the effect of black men's preferences remains uncertain, and it is also unclear to what extent those preferences have been altered by living in a white-dominated society. In some black African societies, there is an explicit preference for overweight women, and waist–hip ratios as low as 0.5. In Mauritania, for example, the desire for larger women is so great that there are 'wife-fattening farms', where women and girls, some as young as seven, are fed high-calorie foods such as dates and couscous, sometimes against their will.

The Hispanic population of the US is usually perceived to have a relatively large body-ideal; indeed, Hispanic women are less likely to feel overweight at any given level of body mass index than non-Hispanic white women. There also seems to be a distinctive peak of body-satisfaction during middle age in this group. In addition, attitudes to women's body shapes show clear inter-generational differences, with older Hispanic people preferring curvier women than younger ones. However, there are also paradoxical trends too – Hispanic women are more likely to be obese, but also more likely to exhibit symptoms of eating disorders, suggesting that they may face even more conflicting pressures than other women.

In East Asian countries, prior to industrialisation, the societal ideal was often relatively heavy, but women in those countries and Asian emigrants to Europe and North America seem to have assimilated the skinnier ideal relatively quickly. Asian-American women's levels of body-dissatisfaction are now at similar levels to those of white women, and one study suggested that Asian men's ideal waist–hip ratio could be as low as 0.6. There remain distinct differences, however. For example, family pressure to reduce weight seems to be a particularly important cause of body-dissatisfaction in East Asian women. Also, it was in this group that a completely new form of eating disorder was discovered – 'non-fat-phobic' anorexia nervosa, in which women under-eat not to become thin, but in response to family or religious pressures, or to prevent perceived symptoms such as nausea and bloating. In fact, non-fat-phobic anorexia has turned out to be a widespread problem in other parts of the world too, including South Asia and the Middle East, as well as the European and North American 'heartlands' of fat-phobic anorexia.

Many sociologists are concerned that the spread of Western media will 'infect' other countries with unhealthily thin female body ideals, and there is indeed evidence to support this tendency. Access to Western television seems to be particularly important in this process – for example, there has been a huge increase in female gym attendance in the Sudan in recent years. Many young gym-going Sudanese women say that one of their main motivations for exercising is a desire to look like Rihanna or Beyoncé, and that they no longer accept the traditions of a country where plumper brides attract plumper dowries.

In contrast, there are some societies which seem relatively immune to the skinny body ideal, even after the arrival of Western media, and some cultures have taken deliberate steps to expunge it. Beauty contests are seen as particularly effective in globalising the Western body ideal, because any woman who wins a small local beauty pageant is then automatically able to 'compete' at ever higher levels up to the global.

A government sponsored group in Burkina Faso actively opposed this creeping onslaught of thinness by running their own beauty contest at the same time as the Miss World pageant. And 'Miss Large Lady' was a great success, eventually won by Carine Rirgendanwa, weighing 117kg, who paraded with other contestants in traditional clothing and bathing suits to win dresses, jewellery and a motorbike.

A few studies have attempted to discern some general rules underlying all this cultural variation, by investigating attitudes to body shape and size across a range of countries. This research suggests that although societies may differ in their preference for women of different body sizes, there is remarkably little difference in their preferences for certain female body *proportions*. This could explain why, for example, the perceived attractiveness of relative measures such as waist–hip ratio is fairly consistent around the world, with just a few minor variations.

A second discovery is that people seem to automatically calibrate their ideas of fatness and thinness to the people around them – for example, someone from Germany will consider the heaviest 20 per cent of Germans to be roughly as overweight as someone from Samoa would consider the heaviest 20 per cent of Samoans, even if one group is actually much heavier than the other.

The third and perhaps most fascinating finding relates to socio-economic status. Studies show that, around the world, people of high socio-economic status tend to express a preference for a slimmer ideal female body shape. People lower down the social hierarchy may prefer either slim or heavier women, depending on the population being surveyed, but the people at the top of the social heap consistently prefer slimness. We will return to this observation in the final chapter.

All this leads to the question of *why* the 'ideal' female body shape varies across the globe and over time. In the preceding chapters we have already seen how female body shape is a crucial factor in human biology and psychology – indeed, it is essential for our survival – so why is something as apparently important as our female body ideal permitted to vary so much?

One simple but often ignored reason for this is that the human species is very physically varied anyway. We are a genetically diverse bunch, living in a wider range of environments than most species, and because of this, different ethnic groups possess different physical traits. Pacific islanders are extremely efficient at storing energy in fat, for instance. East and West Africans have different body shapes and muscle types, and some groups store larger female reserves of buttock and thigh fat. An extreme example of this is steatopygia, a tendency toward extremely large female buttocks and thighs, which has probably evolved multiple times in unrelated human populations – such as the Khoisan and Bantu of Africa, and the Andaman Islanders of the Indian Ocean. These regional variations presumably represent helpful adaptations to the local environment, so a local liking for these characteristics may explain many ethnic preferences for certain body shapes.

Food availability is a second possible cause of geographical variation and could also explain changes in body ideals over time. In a society which undergoes occasional periods of food scarcity, it makes good sense for people to consider larger women attractive – because they stand a better chance of surviving when times are hard. Conversely, by this logic, thin women should be shunned. Indeed, until the twentieth century, slimness was often seen as evidence of parasitism, tuberculosis or other chronic afflictions, and it was only once those diseases had become rare that the current trend for people to be 'skinny, pale and interesting' started.

Yet it still remains difficult to explain why increases in nations' economic development and individuals' socio-economic status seem to lead inexorably to thin body ideals for women. If these links exist, they could explain the shifts towards skinniness in the 1920s, 1960s and perhaps 1990s. On the other side of the economic coin, one study showed that during economic downturns *Playboy* playmates tend to be heavier, taller, older, have higher waist–hip ratios and have smaller breasts. Although changes in social, religious and gender attitudes may play a role, I still suspect that economics is the key.

Our limited data suggest that the female body ideal is 'counter-cyclical' with the economy – when the economy is lean, we like women to be bounteous, and when the economy is bounteous, we like women to be lean. Some have suggested that we each have an inbuilt preference for women who exhibit 'self-control' – who buck the trend and are slim when it is easy to get fat, and who are curvaceous when it is all too easy to become thin. I think there is some truth to this idea, but not that we have a moral preference for women with 'self-control'.

Instead, if we consider the economy to be the modern equivalent of food availability, then it makes sense to prefer women who apparently 'buck the trend'. When times are tough, it seems logical for men to prefer larger women who look like they could bear and support children – and those men's parents should also prefer such women as potential 'daughters-in-law', so this is not all about lust. Conversely, when times are good, it could be argued that men should seek women who can channel all available resources into their offspring, because it does not benefit a father if his co-parent diverts those resources into herself. Thus a good-time girl should be skinny – and in the West times have been relatively good for several decades now.

Whatever the primal causes of the current cult for slimness, there is one aspect of modern life that is often claimed to have been more important than any other in the rapid perpetuation of the skinny ideal: the media. Indeed, some people believe that the media are *solely* responsible for that ideal and blame them for peddling body myths which exacerbate women's body dissatisfaction, stifle gender equality and initiate eating disorders. But what is the evidence?

The artificial visual representation of women began tens of thousands of years ago, if not before. As we have seen, although we know little about the context in which those ancient depictions were created, they certainly vary a great deal in shape and size. The second phase of body-representation did not come until the advent of mass-printing of images in the late nineteenth and early twentieth century in Europe and North America, when for the first time pictures of individual women could be shown to millions of people, sometimes with the implicit message that those women were visually 'superior'. The third phase, and perhaps the pivotal one, was the early twentieth century, which saw the advent of both mass-production-based consumer culture and cinema (and we have already seen that a moving body has a greater impact on the brain than a still one). For the first time in history, women were offered clothes manufactured in 'standard' sizes, and displayed on moving human models. Slim models thus became human coat hangers from which new fabrics and designs could hang artily. The fourth phase came after the Second World War, as television spearheaded the new culture of leisure, lifestyle and self-improvement – and that self-improvement inculcated an urge to match society's ideal body shape. Since then, the media innovations have come thick and fast – a greater focus on women's sport, the cult of celebrity for its own sake, large-scale shameless airbrushing and Photoshopping of still and moving images, social media and the 'selfie', and perhaps most perniciously of all, makeover shows which explicitly claim to improve women by making them conform to the ever-elusive ideal.

Of those eight-or-so phases of media representation of the female body ideal, all except one took place in the last century and a half. This is an astounding acceleration of the rate of change, and we usually assume that this is what created the modern cult of slimness. But do the data stack up to support this claim?

First of all, we need evidence that the media, by their nature, do actually distort female body ideals. Certainly, in television and films made for children, teenagers and adults, studies show that the 'goodies' are usually depicted as slim and beautiful, and the 'baddies' are more likely to be fat and ugly – although some male baddies are permitted to be skeletally thin, just for variety. Women are portrayed as thin more often than men, although this convention may occasionally be subverted so that plot twists can confront us with the horror that a slim beautiful woman may actually turn out to be a baddie after all! Another common media motif is that women with large breasts are presented as being sexually promiscuous, and experimental studies indeed show that people do assume that large-bosomed women are more willing to have sex.

The technicalities of the medium of television are themselves often said to 'add pounds', and actresses and female presenters often comment that they feel they must lose weight before appearing on camera for just this reason. Modern television formats can also be extremely unforgiving – high-definition widescreen television exposes every irregularity of skin tone and texture, while many television viewers do not know how to adjust the aspect ratio on their television, so watch their protagonists either stretched wide or squashed thin.

Printed media and their gawky youthful cousin, the internet, also tend to present a certain image of women's bodies. Studies show that newspapers contain fewer pictures of women than men, and when they do appear women are more often described in terms of their body shape, clothes, state of undress, or who their romantic partner is – and this bias is as true in the sports pages as it is in the headlines. Over the years, women represented in the media have become progressively thinner, and the importance of women controlling their body shape by controlling their eating has been emphasised more and more. This is particularly true of gossip magazines, whose stock in trade seems to be highlighting and commenting on the vagaries of body-shape changes in famous and not-so-famous women – most of the 'newsstand' quotes I listed at the start of this book came from these publications. Fashion magazines also seem to be particularly potent in skewing women's body satisfaction, body ideals and eating behaviour, and no doubt the array of flawless, thin bodies they present is partly responsible for this. I will discuss clothes and fashion in the next chapter, but magazines are where many women encounter underweight, almost boyish fashion models, usually in the context of high-status, high-cost clothes, cosmetics and other miscellaneous products.

Pornography, I suspect, has little adverse effect on the cult of thinness, even though it could certainly exert other negative social effects. Most users of pornography depicting women are still men, although women often have a good idea of the body shapes men secretly ogle. Pornography presents a relatively wide range of female body shapes when compared with other media, and this includes mainstream porn formats – the range of waist–hip ratios of women in *Playboy*, for example, is actually quite wide. Most women are aware that male-directed pornography contains a range of heavier, lighter, taller, shorter, big- and small-breasted, artificially enhanced and as-nature-intended female bodies, and that some of it deliberately focuses on overweight, underweight, pregnant, younger and older women. Whatever the threats of pornography, enforcing the skinny ideal is probably not one of them.

The second line of evidence which links the media to the perpetuation of the skinny ideal is women's anecdotal reports of how media representations of female body shape make them feel. For example, surveys show that teenage girls *believe* that pictures in magazines affect their perception of their own ideal shape, and usually make them want to lose weight. In one study, half admitted that they wished they looked like the models in cosmetic adverts, and a quarter admitted to regularly comparing themselves with magazine images. Interestingly, young women often say that they wish to mimic the magazine-bodies not so much because they want to look beautiful, but more because of the lifestyle that beauty seems to bring – the implied success, glamour, social life, affluence and romantic partners. They say this plants within them the idea that it is important for a woman to 'work' on her appearance to achieve all these things.

Anecdotal evidence like this has its limitations. When women say they believe that an element of the media affects their attitude to their body shape, they may just be plain wrong – they could be giving the answers they think are expected of them, or which they have been 'conditioned' to believe. However, I think it is unlikely that such clear trends could result entirely from error, or 'conditioning' by those who promote an 'anti-media' agenda.

The third line of evidence against the media comes from measurable correlations between media exposure and women's body-shape beliefs. Across the world, exposure to Western, but not local, media correlates strongly with thin-body ideals and women's body-dissatisfaction. For girls and women in the West, although there seems to be no clear correlation between the total amount of media exposure and body-dissatisfaction, there are clear links between exposure to 'appearance-based' media and both dissatisfaction and the tendency to accept a thin ideal body shape. Girls with eating disorders watch more body-related television, especially channels which show continuous music videos. Girls also tend to watch more soap operas than boys, a format which increasingly seems to draw its storylines from sunnier parts of the world where slim young girls often do not seem to wear very much.

Fashion, glamour and 'chat' magazines may show an even greater correlation than television. Women who read these magazines tend to have greater body-dissatisfaction, want to lose weight, and are more likely to have tried to lose weight after reading particular articles. Media habits are changing, however, and women in their twenties spend more time accessing this sort of material online than in print, but here too, time spent viewing appearance-related material online correlates with body-dissatisfaction and disordered eating.

The fourth and final type of evidence comes from studies which attempt to directly link media exposure to women's feelings and behaviour – in childhood, adolescence and adulthood. In this context, dolls may be considered a wing of the media – certainly most mainstream brands of doll come with television and print advertising campaigns, and sometimes there are entire magazines specifically focused on them. A self-contained little doll-world is contrived, a world in which the ideal female shape is almost unattainable. Whereas 2 per cent of men have body shapes which are approximately similar to Ken's (and that, of course, includes me), only 0.001 per cent of women have a body similar to Barbie's – although an internet search for Ukrainian model Valeria Lukyanova shows how a woman can diet, constrain and variously modify her body to become a member of that 0.001 per cent. In one study girls aged between five and eight, especially the younger ones, exhibited a greater discrepancy between their ideal body shape and their perception of their own body after playing with Barbies than after playing with less-skinny dolls, or no dolls at all. In another study, girls ate fewer sweets immediately after playing with Barbies. A Barbie set from the 1960s even included a tiny 'How to Lose Weight' booklet, with the words 'don't eat' written sternly on the back.

Although they may not play with dolls any more, teenage girls' responses to depictions of the female body are not very different. Research shows that viewing fashion websites makes girls feel less positive about their own bodies. More worrying is that other experimental studies suggest that viewing pro-anorexia websites has been shown to make girls who do not themselves suffer from an eating disorder feel worse about their bodies than fashion websites do – one might have hoped that they would resolutely reject the images presented on those sites, and possibly even feel *better* about themselves. Further studies showed that viewing images of extremely fit *thin* women increases teenage girls' body dissatisfaction, whereas viewing extremely fit average-weight women does not.

In general, young adult women exposed to still or video media images of slim women show increased body-dissatisfaction, reduced self-esteem, and more anger and depression – even if those images are computer-generated (*Tomb Raider*'s Lara Croft seems to be a particularly strong stimulus, and casting Angelina Jolie to play her on film probably did not help matters much). However, it is noticeable that the negative effects of these images are not universal. For example, women who start these experiments with high levels of body-satisfaction do not seem to take the same hit to their self-esteem. Indeed, some are *more* body-satisfied after viewing the images, perhaps because they feel that their own bodies compare favourably. Yet still, on average, media images of women's bodies make women feel worse about themselves. One-third of women are less happy about their appearance after reading fashion magazines, and women who read fashion magazines while eating are more likely to stop eating before they have finished their meal.

In recent years, we have seen increasing evidence that makeover programmes are especially potent stimuli. One study showed that women who watched an appearance makeover programme had lower self-esteem and higher awareness of media pressures to be thin than a control group of women who watched a house makeover programme – and that this difference was still detectable two weeks after the media exposure. And because we live during the only time in human history when most humans understand the adverse effects of obesity, media vehicles can now disguise a drive to conform to the societal skinny ideal as a laudable striving to become healthy.

In short, the descriptive, anecdotal, correlation-based and experimental data all strongly suggest that the media are a powerful force in inducing and maintaining a belief that thinness is inherently good, and causing body-dissatisfaction in women who do not feel they have achieved the skinny ideal. As a scientist, I would of course prefer this evidence to be topped off with objective experiments showing exactly which brain regions flare into activity during this media indoctrination process (these experiments are currently underway), as well as studies into whether the tendency to accept externally imposed body ideals is to some extent genetic (these studies are also underway, and the answer seems to be 'yes'). But the evidence we already possess seems very strong.

Yet before we form a mob and burn down the television, magazine and internet companies, perhaps we should consider who the media actually are. The media are created by human beings not too dissimilar to ourselves, who may inhabit an unusual world, but their aim is to make money – and occasionally to inform and entertain too. A good way to make money is to give people what they want (even if they may not yet have realised they want it), so perhaps seeing the media-creators as the evil 'them' and the media-users as the innocent 'us' is simply a way of shirking our responsibility for the monster we have all created. It is tempting – very tempting – to loathe the magazines who have desperately backpedalled from their previous headlines of 'so-and-so shows off her fabulous slim bikini body' to replace them with 'so-and-so shows off her fabulous curves', but people buy those magazines, and they buy them even when they suspect their damaging role in the cult of thinness.

Body ideals vary around the world and over time, but the one we have here in the Western world is currently a skinny one, albeit one filling out some of its curves in recent years. The desire to fit in with that ideal is always strong – to want to be slim, to want a slim partner, slim children and slim friends. Although the ideal may change, there are powerful evolutionary and economic urges to achieve whatever the ideal happens to be at a given time, and it is up to us to resist those urges and to teach our daughters to resist them too.

NINE
Covering up and tucking in

As a matter of fact, our artificial coverings have become so much a part of our life that one may perhaps be allowed to apply the methods of the naturalist to their consideration, and deal with them as if they were part and parcel of the creature which wears them.

The Heritage of Dress, Wilfred Mark Webb, 1908

'Buying bigger jeans feels to me like giving up. I don't know why. I have a real fixation with them. They're my benchmark.'

Anonymous interviewee 'D'

In Florence's Brancacci Chapel hides a true landmark of Western art – a cycle of biblical frescoes painted onto wet plaster around 1424. The paintings' remarkable naturalness and emotional power remain startling today, especially one particular image by Masaccio of Adam and Eve's expulsion from the Garden of Eden. Yet when I visited the chapel many years ago, I was just as struck by the inept way in which later, more prudish artists had attempted to cover the naked figures' genitals with crudely applied fig leaves. Since then, the fresco has been restored and the universal mother and father have regained their former glorious nudity, even though Genesis states that, 'Unto Adam also and to his wife did the Lord God make coats of skins, and clothed them'.

Of all the animals in existence, only humans seem to feel the compulsion to wear clothes – and anything unique to our species is inherently intriguing. Habitual complete nakedness is rare among human societies, yet not only do humans wear clothes, but they often wear far more extensive and elaborate clothes than are necessary simply to perform clothing's basic functions. Our bodies are the aspects of ourselves which we present to the world, and clothes are a dramatic artificial modification of that presentation process. And for women especially, clothes are more about framing, emphasising, enhancing or concealing the shape of the body than the practicalities of keeping warm, and preventing unwanted wobblings and exposures of the flesh.

Of course the clothes women wear, along with other forms of appearance-modification such as makeup, depilation, piercings, tattoos and surgery, reflect the personal *choices* they make to create the 'personal billboard' which I first mentioned in the introduction to this book. And this element of choice is what makes them so interesting. In our quest to find out why women think about their bodies so much, it was important to investigate how those bodies evolved, how they develop, and how they affect health, mind and appetite – yet women do not have much control over those things. To some extent, women must make do with the body in which they happen to exist. Yes, they can decide to change their weight or tone up particular areas, but those things take time and frequently do not work, and surgery is risky and expensive. In contrast, clothes allow us to study what women do when they are able to freely alter their personal billboard – day by day, as they please.

We do not know when clothes were invented, or what form the first clothes took. Some ancient depictions of the human form suggest that ornamentation such as jewellery may have preceded clothes. We also assume that humans had lost most of their body hair before clothes first made an appearance, but again, this is conjecture.

In an attempt to find out when humans first started wear-ing clothes, scientists have turned to a strange quarter for help: lice. Because all lice live their lives among hairs or under clothes, it is thought that human body lice could not evolve until clothes were in widespread use. Body lice probably evolved from head lice, and the two species are certainly similar to look at, yet they each show distinct differences in behaviour – when placed on a human belly, head lice walk up, and body lice walk sideways (a more distantly related third species, the pubic louse, menacingly walks *down*). Thus it is argued that by determining the epoch in which head and body lice went their separate evolutionary ways, we can estimate the time when clothing was invented. Unfortunately, these estimates are inconclusive, putting the divergence of the two species anywhere between 40,000 and 110,000 years ago, and some still claim that clothes may have originated even earlier.

Clothing has three functions – some simple, some complex. It is thought that the first to emerge was concealment of particular parts of the body which were considered 'private' – a concept unknown in other animals. Around the world, those parts most often include the genitals, the bottom, and the female breasts – the regions covered by a bikini, in fact. There are of course cultural variations on this theme: on Brazilian beaches, for example, wearing a thong bikini that exposes the buttocks is considered far more 'mainstream' than on European beaches, whereas going topless is much more frowned upon – and is in fact illegal. At the other extreme, in some Muslim communities, it is believed that *all* of a women's body should be covered, and mesh and wire frames may even be used to cover the eyes. And it was not so long ago that even the exposure of a female ankle was considered risqué in European society.

Clothing may actually be viewed as an enhancement of an ancient tendency for female body concealment to be built into human biology and behaviour. When humans started walking upright, the female genitals became 'hidden' between the thighs in a way completely unlike other primates. This 'vulvo-cryptic' appearance was further enhanced by the fact that the pubic area is one of the few regions of the female human body which still bear hair. Sex is also a much more private activity in humans than in other animals – although one could argue that bipedalism rendered men's genitals *more* exposed, and indeed in some societies the function of male genital coverings is not to conceal the penis but to emphasise it, or even hold it in a mock-erect position.

The human brain can make anything complicated if it wants to, so body concealment has led to some paradoxical effects and it comes with a flip-side: shame and embarrassment at one's nakedness. Indeed, in human societies from hot regions who use body paint instead of clothes, being seen in public without one's paint is considered just as embarrassing as someone in the West appearing without clothes. However, the opposite can also be true, and nakedness is sometimes controversially used as a symbol of power and confidence by women in the West – a badge of their freedom.

A particularly powerful effect of concealment is that heterosexual men find female concealment arousing. Sexual encounters, pornography and men's fantasies all frequently involve an almost ceremonial removal of women's clothes as a prelude to real or imagined sex. In addition, women frequently retain some items of clothing during sex at their partners' request – and of course those clothes may also make them feel sexier, or may reassure them by concealing some perceived bodily imperfection. However, they may also be baffled (or worried) that their paramour prefers them to wear them. Indeed, many men admit that they often find women sexier with their clothes *on* – the photographer David Bailey, best known for photographing beautiful women in the 1960s, once commented that he frequently only became sexually attracted to his previously-naked models once they were leaving his studio in their miniskirt and sweater. Many sexual fetishes also relate to the complex links between clothing or shoes and concealment, exposure, vulnerability, power and pleasure. In humans, sex is uniquely centred within our bizarrely complex (even men's) brains to the extent that the promise of concealment, and delayed gratification, may overwhelm the immediate desire to copulate with a naked woman.

The second function of clothing is the most mundane: clothes and shoes are a good way to protect the skin against cold, heat, ultra-violet light, abrasion, puncture and biting insects. Although protective clothing is sometimes seen as socially and sexually neutral, when it becomes a uniform we can still apply social interpretations to it – think of doctors' white coats, or street-cleaners' reflective jackets. Protective clothing may also become imbued with sexiness if people come to link it to masculinity (for example, firefighters) or femininity (such as old-style nurses' uniforms). If you think I have made this stuff up, visit *uniformdating.com*, or consider how many strippers arrive at the party dressed as accountants or, sad to say, university lecturers.

The third function of clothing is the most complex of all, and will monopolise most of this chapter: the alteration of appearance to signal something to other people. As we will see, women use clothing to emphasise parts of their body, flatter them, augment them, or conversely to hide them. They also use clothes to express their sexual status – their femaleness, maturity and availability may all be independently signalled. For example, a woman in her wedding dress is clearly stating that she is definitely feminine, she is mature yet not *too* mature to have many reproductive years ahead of her, and she is not (and has not been) available for sexual advances by all and sundry.

In the West there is also a fundamental asymmetry between the sexes in that it is often socially acceptable for women to wear men's clothes, but not for men to wear women's. Some suggest this is about appropriation of power or submission, but I am not so sure. Why indeed would most young men be happy for their girlfriend to slide out of bed and pull on their boxer shorts to go and make the coffee, while most young women would react rather differently if their boyfriend wriggled into their thong to do the same thing? The fact that women are often more judged on the basis of their shape and appearance makes it particularly remarkable that in the West it is men who are more constrained by the gender-specificity of clothes.

Expressing social status is also important, and teenagers and adults alike use clothes and fashion to express their membership of particular social cliques. Over time, social groups tend to homogenise their clothing – and this is a big issue for immigrant communities, who must actively decide whether to retain their previous dress style, or express their integration by changing to the prevailing 'indigenous' styles. Clothes can also be used to express formality (people usually laugh when I wear a suit and tie, for some reason), but also social rank and financial wealth – some clothes look 'high-status' and some look expensive, and many people look to fashion brands to send out immediate signals about the cost of their outfit. Status and wealth are not the same, of course – people can be surprisingly negative about others who appear 'nouveau-riche', and some very high-status people deliberately 'dress down', albeit sometimes with a knowing refinement.

As soon as children are born, we start to train them about the social role of clothes – babies are dressed in different colours and different styles according to who their parents want them to be, and who they believe themselves to be. Before the age of ten, children are already acutely aware of the importance of fashion, brands, and the social messages which clothes convey. They understand that they may be accepted, admired, excluded, teased or bullied because of what they wear, and this is especially true for girls.

Despite pressure to conform, children also learn that clothes are a way of expressing individuality – and girls quickly become adept at daily reconstructing their visual appearance according to what they want others to think of them. And that changes all the time – women wear different clothes on different days depending on how they view their body shape. Far more than men, women complain of having 'fat days' and adjust their clothes to their prevailing internal body image. Also, many women use tightness and looseness of clothes as their primary method of assessing their body shape, and studies show that the prime motivation for many women to change their body shape is to fit into certain 'bellwether' clothes. Jeans seem to be especially important in this respect, as my interviewees told me:

I do have some jeans that I can't get into. They are *tiny* jeans, though.

Anonymous interviewee 'A'

It's my clothes that tell me if I've gained weight – it's if my jeans are tight.

Anonymous interviewee 'B'

I've got one pair of jeans that fits, and one pair that's too small at the moment – I keep them because I *will* get back into them.

Anonymous interviewee 'C'

If I'm going out for the night and I put on a pair of jeans and they don't fit, than that's the night ruined. One year my jeans made me cry on New Year's Eve.

Anonymous interviewee 'D'

In other words, clothes are far more than a passive external covering – they become part of their wearer's psychology.

One of the most important ways women use clothes is to alter the visual appearance of their body shape. Every week, millions of words are printed about how various items of clothing can emphasise what women wish to emphasise and conceal what they wish to conceal. However, the basic morphological rules of women's clothing are actually very simple.

I suggest that the female body-to-be-clothed is best viewed as an interconnected set of globose and cylindrical elements. Women instinctively want certain body parts to appear to be approximately spherical, such as breasts and buttocks, but want others to be elongated cylinders, such as legs, arms, neck and torso – and men like them to look like that too. This 'ball and stick' model of female body appearance may seem alarmingly simplistic, but it goes a long way towards explaining why women wear the clothes they do.

First, the globular bits. In earlier chapters, I explained how globular fat-laden buttocks and breasts evolved because they performed useful functions, or because they were sexually selected by men. Women consider them distinctively feminine and men find them sexually attractive, so they are a prime target for sculpting and emphasis by clothes. Because the globular elements of women's bodies are not very structurally resistant to gravity – they sag with age, in other words – clothes often hoist them up to give the impression of youthful pertness. Skirts, jeans and, more recently, underwear and tights are designed with precisely this aim in mind – men's jeans may leave their buttocks looking flat and formless, but women's jeans are designed with structural intent.

For centuries in the West, corsets were the most common way to push bottoms out and breasts up and forward into spherical parodies of their natural appearance, but although cinched waists remain, it is the bra that has now taken over. We once thought that the bra was a relatively recent invention, but that changed with the 2008 discovery of remarkably modern-looking fifteenth-century bras at Lemburg Castle in the Austrian Tyrol. Although moth-eaten, these bras include all the main components of their modern descendants, even including lace trims – suggesting that bras may have supported women's breasts for more of human history than we had previously thought.

As I mentioned in the introduction to this book, human females are unique in requiring mammary support if they wish to run without discomfort. However, most women's breasts are not large enough to contact each other and form a cleavage without artificial support, and here again the bra comes to the rescue. Some modern bras can push the breasts together as well as upwards, and thus can give almost any woman a cleavage – 'Hello Boys,' as the Wonderbra advert once proclaimed, or 'Look me in the eyes and tell me that you love me.'. And of course, bras confer the additional advantage of *preserving* the globular appearance of the youthful breast as the decades tick past, as well as emphasising it on a daily basis.

In my ball-and-stick theory, clothes not only support the globular elements of the body, but they also expose them to varying extents.

Many skirts, trousers and tops contour tightly around the buttocks and breasts, and it has also been fashionable over the centuries to uncover varying expanses of breast. The upper part of the breasts has often been exposed by necklines of differing shapes and profundity, and the French even have a word – *décolletage* – for the region of 'non-breast' exposed above the neckline to give a provocative hint of what lies below. Also, modern advances in clothing technology now mean that women – especially on the Hollywood red carpet – can expose almost every non-areolar part of their breasts if they wish. However, the more the breast is exposed, the less scope there is for support, and these more dramatic feats of mammary exposure might be interpreted as individual women's confident assertion that the breasts involved do not 'need' such support.

The final effect of clothes on women's globular bits is that they define them – they demarcate them from neighbouring regions. Marking the boundaries between globular and cylindrical parts of the body has the effect of emphasising both – for example using necklines to set a boundary between neck and breasts. The lower boundary of the breasts is anatomically more clear anyway, but the upper edge of corsets gives it still greater definition, as does the more recent trend for women to wear opaque bras beneath sheer tops.

The lower boundary of the buttocks is again anatomically marked by an obvious skin fold, but many skirts and trousers tighten at this boundary to more clearly separate the buttocks from the cylindrical legs below. The upper boundary of the buttocks is more vague, which means that women can select from a variety of waistlines to visually define that boundary. Young western women are also increasingly lowering their waistline to expose the upper part of their buttocks, their underwear, or even the upper extremity of their 'intergluteal cleft', in an echo of what they have done with their breasts for centuries. It is striking that young men have also taken to wearing trousers which slip down over their buttocks, although in their case the buttocks usually remain resolutely covered by underwear – perhaps exposed male upper buttocks have too many connotations of crouching low-caste manual labourers. In fact, because men have flatter buttocks their trousers fall down much more easily, whereas the exposure of women's rounded buttocks must by its nature be somewhat deliberate, contrived.

Finally, the ultimate buttock-definers are miniskirts and hotpants – those symbols of confident, modern femaleness. Their upper and lower bounds mark precisely where a women wants her buttocks to be seen to start and end – providing a clear demarcation from cylindrical torso and legs, especially if the midriff and thighs are bare.

The 'sticks' of my ball-and-stick model are the near-cylindrical elements of the female body – the neck, torso, arms and legs.

In Chapter 1 I mentioned that the small female ribcage leaves space for a relatively longer neck, and of course women often deliberately draw attention to it with necklaces, chokers and the necklines of their clothes. Famously, the Kayan women of Burma further lengthen their necks by squashing their collarbones down by the use of multiple large metal neck-rings. The small female thorax, and the absence of men's shoulder muscularity – which probably evolved as an adaptation for throwing in these creatures who are more wedge-slab-and-lump than ball-and-stick – also mean that women's shoulders are small and smooth. However, women more often wear clothing which exposes the shoulders, probably to emphasise just how small and rounded they are, and thus how little space their body occupies. This is, of course, the convention which was entirely subverted in the 1980s when women concealed and extended their small shoulders with shoulder pads and angular polygons of fabric, presumably to assert their confidence in a male-dominated world.

The small female ribcage also leaves space for a distinctively long and cylindrical (albeit flattened front-to-back) abdomen, which can be emphasised by corsetry or outfits in which the midriff is exposed or covered only by transparent material. Women also tend to twist and emphasise their abdominal region when they are flirting. The abdomen is a strangely conflicted area, however, because its long cylindrical form, which denotes youth and slimness, is intermittently replaced by a much more globose pregnant form, signalling fertility and health. No matter how utterly different these two shapes are, both are attractive to women and men, so the abdomen is given special dispensation to alternate between 'ball' and 'stick' configurations, as female biology flips between non-pregnant and pregnant modes.

However, clothing the pregnant abdomen presents particular challenges, and women are often offered a choice between voluminous maternity wear which conceals pregnancy, and shapeless androgynous clothes which defeminize it. In recent years, a realisation that sex, attractiveness and pregnancy are not mutually exclusive has led more women to wear contoured clothes during pregnancy – emphasising their globular abdomen by dint of wearing similar clothes to the ones they wore before they were pregnant. Despite this trend, many people still react negatively to women who *expose* their pregnant belly, unless it is in an 'acceptable and unavoidable' context, such as the beach. The furore surrounding the famous image of a naked pregnant Demi Moore which appeared on a 1991 cover of *Vanity Fair* is testament to the unease which can be induced by a juxtaposition of sexuality, motherhood and exposure.

The arms and legs are the body's most obviously cylindrical components, but I have not discussed them much yet. They are especially cylindrical in women because larger amounts of subcutaneous fat and less muscle mass mean that the contours of women's limbs are more even and straight. Also, due to hormonal differences during childhood and adolescence, women's hands and feet are *relatively* smaller than men's, so their cylindrical limbs are not terminated by the great hammy fists and plodding feet seen in men.

The distinctively smooth, cylindrical nature of female limbs is a major reason why women's clothing exposes more of their arms and legs. Hosiery, especially, is important in enhancing the cylindricality of the legs, and has other functions too – artificially homogenising skin tone, allowing women's legs to remain exposed in cold weather, and toying with exposure and concealment. The popularity of narrow trousers, pencils skirts and long boots shows just how important cylindrical legs are in the expression of femaleness, but there are two further reasons why women emphasise their legs so much.

The first of these is, I think, over-emphasised in the scientific literature. In the West we assume that men are attracted to women with long legs, but the evidence for this is equivocal. Studies show that men prefer female bodies with average, or slightly-longer-than-average legs (the measure used is actually the leg–body ratio). However, this preference is weak, and does not seem to be universal across the globe, so I suspect that it may be the result of conditioning of Western men rather than any deep biological imperative. But the debate continues, and data suggest that women with longer legs are at less risk of obesity, cardiovascular disease, diabetes, liver disease and some cancers. One Chinese study even suggested that they tend to have more children. Thus it remains possible that men could indeed be pre-programmed to seek long-legged women for the genetic and health benefits they possess, and presumably confer on their leggy offspring.

However, I believe the main reason why women's legs are more often exposed than men's is that even if men do not prefer sexual partners with long legs, it makes a great deal of sense for them to choose women with *straight* legs. Having straight legs not only implies that a woman carries good developmental genes, but also shows that she was well nourished as she grew up – presumably by parents who have bequeathed their good child-feeding genes to her. Until surprisingly recently, limb deformities caused by dietary deficiencies were extremely common in the human population, so leg straightness was a highly salient trait for men to be attracted to. Indeed, a historical relic of this obsession survives in the name given to the entire branch of medicine involved with musculoskeletal disorders: 'orthopaedics' means 'straight-legged children'. Thus leg straightness, not length, is the most important contributor to women's attractive limb cylindricality – and it does not matter how light or heavy that woman is. Men have been selected for millions of years to identify straight female legs, no matter how skinny or curvy they are. This urge probably also explains why men find women's stride and gait attractive.

High-heeled shoes are the most common artificial means by which women emphasise their legs, even if those legs themselves are covered. We are often told that women wear high heels to make their bottoms protrude and wiggle alluringly from side to side – think of Marilyn Monroe teetering down the station platform in *Some Like It Hot*. However, the biomechanical evidence for this is poor. As we saw in Chapter 1, a bowed lower back with protruded buttocks is indeed a distinctively human characteristic especially pronounced in women, but there is no evidence that high heels make women bend their lower spines forward or stick their bottoms out more. If anything, research suggests that they bow their spines *less*. Similarly, while women exhibit more swinging and rotational movement of their pelvis as they walk, kinematic studies show that high heels do not increase those movements, although they may induce a very small 'rocking back-wards and forwards' movement in the pelvis.

Instead, I suggest that women wear heels for completely different reasons. First, they force them to walk slowly, and with shorter steps, thus emphasising those two characteristic features of female loco-motion. One biomechanical study demonstrated that women walk like this to compensate for the instability inherent in high heels – they are not trying to walk in an ultra-feminine way: they are simply trying not to fall over. The second reason for wearing high heels is that tilting the foot makes it take up less horizontal space, thus creating the illusion that it is smaller. Relatively small feet are also a distinctive feature of women, and have been shown to be attractive to men in many different cultures around the world – indeed, some cultures resort to foot-binding to artificially shorten girls' feet. The third reason for wearing high-heeled shoes is that the foot is tilted to become more vertical. This creates a further illusion that the foot has been somehow incorporated into the cylindrical portion of the limb, making it appear longer (which may or may not be desirable to men) and straighter (which definitely is). The same applies to long gloves, which make the hands appear smaller by 'incorporating' them into those alluring arm-cylinders.

Women put up with a lot when they wear heels. They alter their gait so that hips and knees must bend and straighten more; there is more sideways movement at the knee and ankle; the foot rotates differently when it strikes the ground; and there is more stress on the knee, which may cause arthritis in later life. Muscle activity is greater, leading to rapid fatigue, and venous blood ascends the legs more sluggishly. It is believed that the brain has to completely rewire its locomotor control systems to cope with high-heeled shoes, and this could have knock-on effects on other brain processes – one rather far-fetched scientific paper even suggested that wearing high-heeled shoes is linked to schizophrenia. And in a triumph of excessive accuracy, it has been calculated that the average pair of high heels causes pain after being worn for 66 minutes and 48 seconds. In another study a third of women admitted to dancing barefoot at parties because their high heels hurt too much, and a third admitted to having walked home barefoot for the same reason.

Yet the urge to meld the foot into the leg-cylinder must be strong because one-third of women also admit to buying high-heeled shoes which are too small for them, simply because they are beautiful. And indeed, high-heeled shoes are often beautiful objects in themselves, and the more expensive they are, the more likely they are to be presented like fetish-objects in smart boutiques. The average Western woman owns more than twice as many pairs of shoes as the average man, and they certainly seem to be comfort items – one study showed that women who react with greater insecurity to images of attractive female bodies tend to own more shoes.

All in all, the ball-and-stick model of clothing the female body goes a long way towards explaining how and why women use clothes to emphasise, augment or conceal their body shapes. Yet one further aspect of clothing which often irritates and mystifies women is why fashion models are so skinny – why they are all stick and no balls, as it were. Modelling clothes on exceptionally slim women could be argued to be counterproductive – it makes it difficult for most women to imagine what they themselves would look like in those clothes, and it probably makes many assume that the clothes are 'not for them'.

Of course, stick-thin models could all be the result of the old problem, the cult of slimness, although that would not explain why fashion models are usually thinner than actresses or television presenters. However, fashion models are employed to sell fabric, not bodies, stories or information, and much of the imagery of fashion is akin to 'fabric porn', with material flowing, floating and billowing as models move, often in filmic slow-motion. And thin models are like human coat hangers from which fabrics – the product – can hang unimpeded, with no distracting curves on which to get snagged. Fashion is about advertising and like all advertising its imagery – the clothes, the bodies and the lifestyle it peddles – is somewhat removed from the real world.

In addition to manipulating the ball-and-stick system, women can also alter the appearance of the shape of their bodies by exploiting the colours and patterns of fabric. Studies show that women tend to feel more strongly about the colours of clothes than men do, and there seem to be differences – either innate or learnt – between colour preferences in the two sexes. In surveys, women are more likely to express a preference for particular favourite colours, although those favourites are more likely to be pale and pastel shades, than the simple bright colours which men prefer. However, colour is important to both sexes, and in one study in which people were asked to select their favourites from a series of pictures of women wearing various outfits, the most common reason for liking or disliking the clothes was their colour. There also seemed to be a remarkable consensus between the experimental subjects and a panel of fashion 'experts' as to which colours looked best.

Even before they can speak, there is evidence that babies prefer red, blue and purple, yet dislike pink. There has been a great deal of argument about why little girls get dressed in pink and little boys get dressed in blue, and this issue is important because gender-specific colouring persists throughout life. Most theories have focused on the link between blue clothes and the greater 'value' of male children – blue pigments made from minerals such as lapis lazuli or azurite were historically more rare and expensive than pinks, and perhaps even considered 'regal'. Blue was the colour of the sky and spirituality, so blue may have been thought to protect valuable boy-children from danger (although families throughout history have dressed their young sons like girls to trick evil spirits into not considering them worth harming).

I suggest that dressing girls in pink is actually the first stage in the deliberate packaging of the Caucasian female body as something to be exposed and viewed. Pink is, obviously, very close to Caucasian skin colour – it is almost like not wearing clothes at all. And as soon as toddlers wear clothes, little girls are more likely to expose their legs and arms than little boys. As we have seen, this continues into adulthood, with women's clothing exposing greater expanses of leg, arm, neck, back, chest and midriff than men's. Women's clothes enhance this sense of exposure with lace and other semi-opaque materials rarely seen in men's clothes – indeed, men's clothes are usually distinct, blocky shapes, with clear edges and boundaries which emphasise their 'wedge-slab-and-lump' body shape. So pinkness is just another way in which the female body is made to inhabit the marginal zone between being clothed and exposed. And 'nude' fabrics and shoes, which have become increasingly popular in recent years, take this process to its logical conclusion, by clothing women in a colour explicitly designed to look like it is not even there.

It seems rather hypocritical that women are told to cover their modesty, yet continually pretend to expose themselves. However, pink and red may also have deeper biological and psychological effects. Most mammalian species can discern colour – pet cats and dogs can, hoofed animals can, and fruit bats are probably just as colour-discriminating as we are. In many species the colour red is attractive to males – so red is often used as a sexual lure by females. It may not seem likely that this phenomenon could extend to humans, but studies show that wearing red makes men see women as more sexually desirable, although not more kind, intelligent or likeable – it is, after all the colour most associated with the vamp or *femme fatale*. Conversely, wearing red does not seem to affect women's perceptions of each other. But when women were asked to select a picture of themselves to put on their web profile, those who had previously said they were interested in finding a sexual partner were statistically more likely to choose an image in which they were wearing red. Similarly, women selecting pictures to place on a dating website oriented towards relatively casual sexual relationships were more likely to choose a red- or pink-clad picture than women registering for a dating website focusing more on long-term relationships.

In fact, there is increasing evidence that women use coloured clothing to send men signals about sex and fertility. Humans are exceptional among mammals in that women do not exhibit obvious phases of oestrus or 'heat' – distinct time-windows of sexual receptivity around the time of ovulation – almost *all* other animals do, including our closest relatives, the chimp and gorilla. Thus human males are effectively kept in the dark about female fertility. Indeed, there is good evidence that non-scientific human societies do not actually know when women are most fertile – the ancient Greeks thought women conceived when menstruating, and present-day Hadza hunter-gatherers still believe that women are most fertile immediately after menstruation.

However, women do seem to use clothing – largely unconsciously – as a way to give men some heavy visual hints about their fertility. For example, men rate women's clothing choices as more attractive at times of peak fertility – and cyclic fluctuations in the perceived attractiveness of clothing are greater in single women. Women's oestrogen levels correlate strongly with the tightness of their clothes and how much skin they expose at social events, and their stated desire for sex correlates with the sheerness of their clothes. Finally, women are three times more likely to wear red or pink around the time of ovulation.

But it's not just red that sends out signals: white and black also speak loud and clear. White clothes provide a very clear body outline, and contrast strongly with dark and tanned skin, and it is for this reason that women tend to wear white when they are feeling confident. However, black is the most frequently worn colour of all, partly because it does not clash with anything, but also because it serves other, sometimes contradictory purposes. During the day it provides a bold outline, while at night it does the opposite, making the body seem smaller. Also, although it provides strong contrast with pale skin, a predominantly black outfit provides almost no contrast across the body, so women can use it to conceal perceived body-shape imperfections. This double-nature of black probably explains why it is so popular, and especially why the 'little black dress' has become almost a uniform of smart, sexy femininity – it reduces the apparent size of the body, emphasises the globularity of the buttocks, accentuates and exposes the cylindricality of the limbs, exposes an acceptable amount of cleavage, yet cannot look chromatically gaudy.

Indeed, women anecdotally report that they feel better about themselves when they wear red and black, and this may have un-expected effects. In one study, women were photographed wearing different colours, and then those images were cropped so that their clothes were not visible. When other people were asked to rate their attractiveness, women wearing red and black were considered more attractive, *even when their clothes could not be seen* – so presumably wearing those colours makes women appear visibly relaxed, confident or happy. Wearing black and red really does seem to make women 'act beautiful'.

Women are also not averse to using optical illusions to alter their apparent body shape. Perhaps the most simple example of this is the recent trend for dresses to include elongated triangular panels of sharply contrasting dark fabric which 'indent' the visual appearance of the sides of the body to make the waist look narrow. Yet the most common way in which women attempt to fool the casual observer is by wearing clothes with vertical stripes in the belief that this will make them look taller or thinner. This is actually based on a complete misunderstanding of what is called the Helmholz illusion. In fact, the nineteenth century scientist Hermann von Helmholz actually noted that vertical stripes make an object appear shorter and squatter than it is, whereas horizontal lines make it look taller and thinner. The illusion is a very real one, and works with bodies just as well as with other shapes. It is unclear why the fashion industry got it so utterly wrong.

Beyond clothing, there are several other forms of female self-alteration, ranging from the mild to the unnatural or invasive.

The first of these is skin-colouring. Even more than the clothing industry, the cosmetics industry aggressively promotes the idea of a visual ideal to which women should conform, and that to conform they must conceal their blemishes. Cosmetics companies frequently use phrases like 'airbrush' and 'age rescue' in the names of their products. And, whether or not the science behind these claims is plausible, it seems that wearing cosmetics does consistently improve women's assessments of their own attractiveness – it is no wonder that many women refer to their makeup as war-paint. Makeup makes men rate pictures of women as more attractive too, even if they are exposed to those pictures for as little as one-quarter of a second.

One of the main functions of makeup – for the face, but increasingly for the body – is to homogenize skin tone, something which we instinctively associate with health, youth and stubble-less femininity. Also, eye-tracking studies show that people's gaze dwells longer on faces with an even skin tone, even though there is actually less surface detail to look at – the observer basks in a reflected featureless glow.

Modifying skin colour transmits several different messages. Paleness is seen as an attractive ideal for women in most human societies, suggesting that it is a preference which has deep biological and evolutionary roots. Indeed, a whole skin-whitening industry has developed, focused on darker-skinned ethnic groups. Paleness is probably a strong psychological indicator of femaleness, because male sex hormones tend to darken the skin. Indeed, this male desire for paleness has been claimed to explain the surprising speed with which human populations emigrating from sun-drenched Africa lost their skin pigment. Genetic studies suggest that Europeans and East Asians independently and rapidly lost much of their skin melanin, despite the fact that it still confers considerable protection from the sun even in temperate zones – and the speed of this pigment loss may be explained by sexual selection of pale mates by men. It was once thought that this process also explained the evolution of blond hair, although there is no evidence that men have an inbuilt preference for blondes, whatever Marilyn might have claimed. Only 2 per cent of the world's women are blonde, and studies suggest that men are attracted to blondeness simply because it is unusual.

Pale skin also creates a greater contrast with facial features, nipples, pubic hair and genitals, and this contrast is in itself attractive to men. Skin contrast is greater in unmade-up women than in men, and makeup further enhances this feminine contrast by darkening eyes and mouth, and making skin paler. For example, computer-generated images of androgynous people can be made to appear more feminine or masculine to independent observers simply by increasing or decreasing their colour-contrast. Contrast is also enhanced by reddening the lips to mimic the effects of oestrogen or sexual arousal. In this context, the cheekbones seem to qualify as facial features too, as they are also often blushed with an artificial mimic of a womanly oestrogenic glow to set them apart from the surrounding skin.

If pallor is universally attractive, then it seems strange that suntans go through phases of being considered beautiful. However, tanning is not considered desirable in most human cultures, and even in the West it was viewed as a sign of a life of low-prestige outdoor labour until relatively recently. With the advent of Riviera holidays and Mediterranean cruises in the 1920s, tans became associated with affluence, leisure and the jet-set lifestyle – and this is presumably why cheap, inept artificial tans entirely negate tanning's allure. I would also argue that tanning temporarily homogenises the appearance of the skin, and that the gradations of tan between exposed and less-exposed areas accentuate the smooth cylindrical appearance of the limbs, thus making the body look younger and thinner. Of course, long-term sun exposure decreases skin tone evenness and increases wrinkling, yet the drive to look glowing and well-travelled in our youth often seems to override most concerns about later-life leatheriness.

Along with skin-colouring, depilation is one of the most common body alterations, although it does not really affect body *shape*. Surveys in developed countries suggest that up to 99 per cent of women remove some sort of body hair – perhaps 96 per cent remove leg and armpit hair, 60 per cent remove at least marginal strips of pubic hair, and among those, 48 per cent remove most pubic hair. There is evidence from the graphic arts that female depilation has a long history – in Western art few pre-twentieth-century female nudes have pubic hair, for example. The aim of most non-pubic depilation is probably to enhance women's relative hairlessness – or rather the fact that much of their hair is the almost-invisible 'vellus' kind, instead of the coarse, pigmented 'terminal' hairs which cover men. In contrast, there is evidence that pubic hair removal is more related to sex. In many cultures it is associated particularly with the sex industry, but the now widespread trend for increasing depilation demonstrates the tendency, first noticed by the Romans, for the proclivities of sex workers to gradually work their way into the mainstream. In the West, pubic hair removal is more common in young, sexually active women, some of whom claim it improves the physical experience of sex. Some women say they depilate because it makes them feel clean, and others because it makes them feel young. This leads to a potential conflict which may explain why complete pubic hair removal is less common – women may want to look young, but they do not want to look like children.

Moving further along the 'unnaturalness' spectrum of body alteration, we come to piercing, tattoos and scarification. These may in fact be the oldest body alterations of all – for example, a 5,000-year-old man found frozen in ice in the Austrian Tyrol had very distinct and elaborate tattoos. More than any other form of body adornment, tattoos are linked to initiations into particular groups, expressions of status, and reclamation of power over one's own body. Certainly, they are found all around the world – the word 'tattoo' itself is probably derived from Tahitian. They can be signals of defiance or spirituality, record milestones in one's life (including drunkenly stumbling into a tattoo parlour, presumably), but are increasingly used by women to enhance their appearance.

Most women's tattoos come in relatively few formats – the first of which are usually no more than small, often hideable, badges of femininity, individuality or love – a name, symbol, or a figurative or abstract design. The second type consists of sinuous or floral designs twining around the arm, ankle or even the torso, presumably to emphasise the slim, cylindrical nature of those body parts. The third is the increasingly common 'tramp stamp' – a variety of horizontal motifs etched just above the bottom, as a permanent demarcation between the straight back and the globose buttocks.

Piercings are much more common in women than men, and the almost ubiquitous earlobe piercings are now often accompanied by piercings of the ear cartilages, lips, tongue, nose, nipples, navel, clitoral hood, and almost any other piece of skin which comes to hand. Tongue and genital piercings often have sexual symbolism or uses, but most other piercings enhance the perception of body shape by contrasting their physical hardness and metallic 'foreignness' with the most distinctively soft and vulnerable parts of the female body – necks, lips, noses, bellies and breasts. The more tender the area, the more arresting the piercing.

Finally, the most extreme form of body-shape repackaging is surgery – which shows just how far women will go to change their own and others' perceptions of their body shape. In the US, and probably most developed countries, the most common cosmetic surgical procedure is breast augmentation and reshaping. The most frequent motivation for breast augmentation is usually women's own body dissatisfaction, and it seems that male sexual partners rarely suggest the procedure. Most women who undergo breast augmentation are young and married without children, although the demographic range is wide. Although most surgeons would prefer to create enlarged breasts which retain the natural, pendulous teardrop shape, many women ask for their upper breast to be especially enlarged, to give the bosom an unnaturally spherical shape and provide a dramatic cleavage in plunging necklines – a sign not only of the importance of clothes in body image, but also how the 'ball-and-stick' urge can get out of hand.

Almost as common is liposuction. Women more often wish to lose fat from their thighs than from their bellies – although there are few data regarding how often men actually want their partners' adipose deposits to be reduced in this way. Next is the abdominoplasty or 'tummy tuck', which is often carried out following dramatic weight loss, or to expunge the effects of pregnancy on the abdominal skin. And, following on from the popularity of breast augmentation, the use of artificial implants to expand other female curves such as the buttocks is, although infrequent at present, rapidly becoming more popular.

Surgery to change the shape of the body is inherently controversial. Many cultures believe that the body is a heaven-sent thing which should not be unnaturally altered, and many non-religious people feel the same way. Plastic surgery was originally developed to help people with injuries and disfigurements, but the idea of surgery as a valid way to enhance healthy bodies has progressively taken hold over the last century. Surgical augmentation is, by its nature, something which women can take or leave as they wish, but in this new century that view of cosmetic surgery is being overtaken. Many fear that surgery is becoming normalised – an accepted way not just to augment a healthy body, but also to achieve acceptance and conformity by excising perceived flaws, and gain membership of an exclusive club of women sculpted in distinctive, recognisable ways by particular surgeon-designers. A gradation of procedures now exists in which botox and fillers are seen as innocuous, almost makeup-like activities, but which may act as 'gateway' procedures – the cannabis of cosmetic alteration, leading inexorably to ever more invasive and irreversible surgeries. Women seeking surgery – an ordeal once only countenanced for the most severe and acute illnesses – now often see it as a way to increase their control over their lives and to hold on to their youthful allure.

Controlling the appearance of women's body shapes – redesigning the personal billboard – has become a dominant force in our society. Each year in China a beauty contest takes place which women may only enter if they have undergone a cosmetic surgical procedure – and they must submit written medical evidence as proof of their surgery. On the other side of the world, the clothing company Abercrombie and Fitch have provoked criticism for offering men's clothes up to size XXL, but women's clothes only up to L, in response to which their CEO Mike Jeffries said that he did not want 'fat and ugly' people wearing his company's clothes, let alone working in his shops.

We are continually told that the ideal body is something to which women must aspire yet never reach, at enormous financial and emotional cost. And the elaboration of the body – connecting up those balls and sticks, all that homogenisation, plucking and puncturing – is the most overt expression of that idea. Being able to alter one's body shape may seem like the ultimate in choice, but really there is not much choice at all. Women are encouraged to be visually and aesthetically individualistic, but only if they are individualistic like everyone else.

Body shape can seem like an internal, smouldering furnace of unease and dissatisfaction at the centre of modern women's lives, yet as we near the end of this book, I still have not fully explained why they *care* so much about it. In my introduction I said that my quest into female body shape was driven by two interlocked questions. We have explored the first question – why human females have such a special body shape, and how that strange shape feeds into almost every area of their lives. So it is now time to draw together everything that has gone before, and answer the second question. It is time to find out why women think about their bodies so very much.

TEN

Why women care and why it's complicated

Come, you spirits
That tend on mortal thoughts, unsex me here,
And fill me from the crown to the toe top-full
Of direst cruelty!

Lady Macbeth, in *Macbeth*, William Shakespeare, c.1606

'Women judge other women more than men judge other men. Women are mean and bitchy. I do think the way I look is important to me. I feel more pressured to look good if I'm out with my girl friends than with my partner. I want them to see me as attractive – I want to seem at least on a par with them. I guess there's an instinct – you want to look better than everyone else. It's always nice to think you've got a better body than the people close to you, isn't it?'

Anonymous interviewee 'A'

In the last few million years humans have acquired a truly remarkable repertoire of unusual adaptations – a combination of exceptional and often unique characteristics never before seen on earth. We walk on two legs, breed in bizarre ways, use tools and change our environment – and most of our success has resulted from the unparalleled cognitive abilities of our vast brains.

Yet that oversized brain consumes enormous amounts of energy, and throughout pregnancy and much of childhood, most of that energy is provided by mothers. The obligation to nourish developing brains has come to dominate our entire species, and women's bodies in particular, and as a result a great deal of human life has become fixated on that uniquely curvaceous female form. Indeed, so great are the biological, psychological and cultural influences of the female body, that they have now been distilled down to create the central element of the most complex human arena of all – our social world. So now we must take a final journey through the strange social world of female bodies, and finally discover why they obsess us so.

Competition

The relationship between female and male humans is unusual. Compared to other species, even our close relatives such as gorillas and chimps, men are not large, overwhelmingly dominant or promiscuous, yet the human sexes are not very similar, either. Both sexes inherited the vast brain that goes with humanness, but their inescapably different contributions to the next generation gave them different bodies. No matter how much we might wish to escape from the notion of biological determinism, evolution is not feminist.

Women's bodies are all about storing calories for future offspring, and being *seen* to store those calories, whereas men's bodies are focused on cooperative acquisition of yet more calories once those offspring have been conceived and born. In all human societies there is a division of labour between the sexes, and while that division varies from culture to culture, the end result is that children are usually sustained by two parents, but do not have to accompany those parents on perilous foraging expeditions. Parental cooperation and division of labour were a human innovation which meant that hunter-gatherer children were, in short, less likely to die than other young primates.

Because human children proved to do much better when both parents invested in them, the human sexes gradually became realigned. For example, it now makes sense for women to seek men who are not only genetically superior, but are also likely to be good providers. Conversely, because men have to help out with their offspring, it behoves them to think very hard about whom they have those offspring with. Just as males of most species must compete for females, so also must human females compete for a limited pool of males to sire and support their babies. In a promiscuous or polygamous species, any female can pretty much count on being mated by a superior male, but in humans, infant-provisioning males are a valuable commodity and they must be competed for.

Although humans are quite unusual in this respect, female within-sex competition has actually been studied in a variety of non-human species, including elephants and non-human primates. Like males, females compete for resources, and in species like ours they must also compete for mates. However, their modes of competition are often more subtle than those of males. Instead of clashing antlers or baring teeth, females emit subtle signals of aggression, territoriality and dominance which influence the behaviour of other females. The presence of a dominant female can even suppress the fertility of other females – a phenomenon seen in baboons, talapoin monkeys and naked mole-rats. Thus female within-sex competition is hard to study because it is not overt, although it is nonetheless real. And, as we would predict, monogamy increases its ferocity. When desirable males are in limited supply, females fight hard for those 'few good men'.

Many female primates, including chimps, macaques, baboons and vervet monkeys arrange themselves into dominance hierarchies – pecking orders – and these hierarchies dictate who gets to eat more, who produces more babies and whose babies survive. Not only is female dominance genetically inherited from stroppy parents, but dominant mothers can also intervene directly to support their daughters during confrontations, thus further ensuring their future dominance. Most animals actively seek to achieve high social status, and studies show that high status reduces quantifiable signs of stress, while low status increases them. And the same hierarchies exist in humans too – with low status correlating strongly with stress, unhappiness and feelings of defeat and entrapment.

The establishment of the main two, largely independent, dominance hierarchies in humans – male and female – takes time. Children already express preferences for certain individual peers by the age of three, and hierarchical interactions are evident soon after. Male and female children form a rather messy unisex hierarchy at first. Boys tend to be more dominant than girls, although there are many exceptions to this, and children's dominance may also depend on context – children who are submissive at school may be dominant at home, for example. Young boys use assertiveness and aggression more than girls, and compete particularly pushily with girls at sport, yet male aggression is rarely a long-term strategy for maintaining dominance. Indeed, in newly concocted social groupings of children, boys' competitive interactions wane with time, whereas girls' slowly increase. Interactions with parents differ too – by mid-childhood both boys and girls tend to see their same-sex parent as more dominant and likely to punish them.

The separate male and female hierarchies become more clearly established in the teenage years. Boys compete more exclusively with boys, and girls with girls. Many teenagers notice that, around the start of secondary school, their social world changes from being a set of friend–friend pairs to a looser arrangement of social interactions. Throughout the second decade of life, a more adult-like arrangement of dominance hierarchies emerges – two major heterosexual pecking orders defined by sex, in addition to other dominance relationships between homosexual teenagers. Girls and boys all slowly become less assertive to members of the opposite sex and more assertive towards members of their own sex.

In adult life, within-sex social status is a potent force. For example, studies show that people who are the subject of adverse comments by members of their own sex are less likely to find romantic and sexual partners, and to have children. And, women who are seen with high-status men are assumed themselves to be of higher status by other women, but not by other men. In addition, there is considerable evidence from studies of small groups of co-workers that women express particularly negative perceptions of female authority figures, and that this has adverse effects on their mood. You have a place in the hierarchy, whether you like it or not.

Aggression

Girls and women establish their pecking order in a distinctive way. While boys and men are flexing their muscles and gamely beating the hell out of each other, the female of the species focuses on the verbal.

From as early as the age of two, if girls are put in a room with other girls they do not know, they find more to talk about and experience less embarrassment than boys in the same situation. Rather than discussing abstract concepts, or even talking about themselves, girls' conversations often tend to discuss relationships – especially relationships between themselves and other girls. As boys get older they tend to talk more about their interests or possessions, and argue more, whereas girls develop 'girl-talk': prolonged and complex discussions of people, social norms and their beliefs.

Another characteristically female form of discourse which develops in late childhood is 'fat-talk' – frequent and sometimes repetitive commentaries on one's own fatness and body-shape defects. The function of fat-talk seems to be to establish girls' places in their circle of friends by asserting their awareness of their body, and establishing a socially acceptable level of confidence and contentment about it. Thus by the time girls start secondary school, body shape is already a tool to establish social status.

It is usually socially unacceptable for girls and women to be overtly physically aggressive, and indeed it does not make evolutionary sense for members of the sex which usually provides most parental investment to get badly injured during physical conflict. Because of these restrictions, women more often use 'indirect aggression', usually verbal, to establish their position within the dominance hierarchy. The aim of indirect aggression is to damage other women's relationships with their peers, but it also has the advantage that it can subsequently be denied – either because nothing explicit has been said, or because a woman's friends can 'close ranks' around her if things get out of hand. And casual verbal aggression still allows women to maintain a degree of social 'presence', or poise – an ability to control one's emotional reactions and be sensitive to social cues which is itself strongly correlated with high social status.

Research indicates that the most common modes of indirect aggression are gossip, insults, shunning, ostracising, stigmatising, accusations of promiscuity and, notably, giving advice. We often assume that giving advice is an altruistic rather than a dominating activity, yet women higher in the social order are far more likely to provide advice and women lower down are far more likely to receive it. Women also occupy much more of their time with gossip than men, they encourage each other to gossip more, and they gossip much more about close friends and about other women's appearance, often questioning the attractiveness or dominance of other women.

The importance of indirect aggression may explain the otherwise confusing fact that cliques of girls or women who are thought of as 'mean' – nasty to other women – may nonetheless be popular. Indeed, some studies suggest that 'meanness' is strongly linked to popularity – so maybe dominant women use meanness as a form of indirect aggression to maintain their dominance. Another phenomenon which can spill over into indirect aggression is sexual jealousy. This tendency of humans to 'guard' their mates from marauding suitors is just as evident in women as it is in men, such is males' parental value in our species. A well-characterised extension of this mate-guarding is women's social intolerance of other women whom they consider to be excessively sexualised or provocatively dressed – an intolerance which may sometimes lead to individuals being permanently excluded from social groups.

There is a huge amount of psychological data showing that women exist within a tumult of social competition, fuelled by an indirect verbal aggression which contrasts strongly with the good old violence which makes men's lives so straightforward. And the evidence strongly suggests that indirect aggression *works*. The recipients of within-sex indirect aggression tend to start dating later in life, they flirt less, they find it harder to find romantic and sexual partners, and their romantic relationships do not last as long.

In other words, being socially dominant is enormously important for a woman.

Appearance

And this is where bodily appearance becomes all-important.

In women's dominance hierarchies, the primary determinant of a woman's social status seems to be her visual attractiveness. Many studies have investigated the factors which allow individuals to clamber to the top of the league, and they usually agree on the criteria for success. And a depressing bunch those criteria are, to be sure. For men, physical strength and prowess seem to be important earlier on in life, and wealth becomes important later on, whereas for women, attractiveness comes out above intelligence, athleticism and social ability right across the age range. This starts early – even before puberty, attractiveness already seems to be more important to girls' status than their abilities or reputation. The primacy of female attractiveness is a common thread which runs through different human societies around the world. Wherever one looks, attractiveness seems to be the route to the top of the social pile.

And of course the compulsion to reach the top of that pile explains why women think about their physical appearance so much. Women are not directly competing for men – I think many of us instinctively suspected that was the case – but they are competing for status, which presumably then *indirectly* gives them the power to choose the mate they want. Competing for competing's sake, almost.

This urge to compete explains why young women report that they experience far more episodes of appearance-based within-sex competition than men. Women are also more interested in what other women think about their appearance than what men think – which probably explains why my wife ignores my opinions of her clothing choices in favour of those of our fifteen-year-old daughter. Informal surveys suggest that one in ten women openly admits that she always wants to look 'better than the competition', although four in ten said they compared themselves to other women and the same proportion said they tried to compete with other women or impress women more than men. Roughly half admitted to complimenting women they did not know on their appearance, whereas almost none would compliment a man whom they did not know. Women also believe women's compliments about their appearance more than men's – although that may say more about men than it does about women.

Other studies show that women's competitiveness is complex and context-dependent. For example, one demonstrated that women are more attracted to men who already have attractive female partners. A woman's dominance can also, apparently, transfer to the man she is with – and men too seem to understand this innately. Studies show that men with attractive female partners are more likely to choose social situations in which they can mix with people of their own age, to 'show off' their partner to their peers and competitors, while men whose partners are less attractive seek out older social groupings.

Women's beauty is so socially important that it has become embedded in the way we all judge women. When shown a series of images of women and asked to guess what sort of characteristics those women may possess, experimental subjects tend to assume that attractive women are more likely to be honest, well adjusted, happy, employed in prestigious occupations, sexually experienced, happily married, sociable and, most of all, popular. They even assume that attractive women are better at performing a variety of tasks.

Of course, the effect of others' assumptions is that beautiful women are treated differently all their life, and there is strong evidence that they behave differently as a result – more 'beautifully', perhaps. Studies show, for example, that attractive women communicate with more animation and visible enthusiasm than women who have been scored as less attractive by independent assessors. Our assumptions about beauty become a self-fulfilling prophecy.

People are just plain nicer to attractive women, and this seems to be the case even if they do not even expect to meet them. In one study, bogus application forms for a university graduate course, complete with a photograph of the bogus applicant, were left in public telephone boxes as if mislaid by those applicants – and the forms with photographs of attractive people were far more likely to be forwarded to their intended destination by passers-by than forms emblazoned with photos of less attractive people. And this appears to be a general phenomenon – people are more likely to help, employ, give a reduced prison sentence to, want to be friends with, or admit to university, attractive women with high status.

However, there are two complicating factors in all this beauty-bias. The first of these is that women's competitiveness is a multi-layered thing. For example, although women usually assume that beautiful women are more successful, they are also more likely to attribute their success to luck than to ability – whereas they are happy to assume that unattractive women or attractive men's success is due to their ability. The second complication is that being attractive changes women's perceptions of themselves. Thus women who are considered attractive by others, themselves tend to prefer men who are more symmetrical, masculine and attractive. One experiment showed that receiving complimentary or derogatory (fake) comments about one's appearance makes people select respectively more or less attractive opposite-sex partners with whom to carry out a subsequent task – and the effect of the nice or nasty comments is significantly greater on women's choice of partner than on men's.

The importance of visual appearance in the dominance hierarchy fits well with evolutionary and reproductive theory – for two good reasons. First, it makes sense for women to compete for social status and a limited supply of mates by emphasising the things that men find attractive – and, as we saw in Chapter 4, visual appearance is indeed extremely important in men's choice of partners. This is also backed up by the finding that men are considerably less picky about visual attractiveness when selecting a partner for a one-night stand than when they are picking a life-partner.

The second reason is that it would make sense for women to compete in terms of visual appearance if, and only if, appearance is a true indicator of their genetic fitness. And as we have seen previously, there is a great deal of evidence that female physical attractiveness – especially the body-shape-related characteristics which men seek – has a strong correlation to genetic and physical health. From this perspective it is obvious *why* visual appearance dictates social standing: over the course of human history, attractiveness has been an accurate indicator of future potential as partner and parent.

Within-sex competition probably also helps to explain the accentuated curviness of human females and the dramatic elaboration of women's clothing. In many species including deer, roach, flies and birds of paradise, within-sex competition for mates evolves alongside increased body ornamentation – and in humans that would mean females focusing resources on looking good because they have more reason to compete. This may be why sexual selection has acted so strongly on human female body shape in the past, and why women today put so much effort into augmenting and emphasising their appearance. A further refinement of this tendency is that women wear different, and apparently more attractive, clothes around the time of ovulation, when they are most likely to conceive.

A picture is emerging, and perhaps a slightly disheartening picture, which places visual appearance at the centre of human female life. Linking female appearance, attractiveness and status now seem somewhat unpalatable – perhaps we thought we had left all that stuff behind, somewhere in the 1970s. However, I must once again emphasise what I said at the start of this book – each of us is the product of a sequence of thousands of heterosexual couplings by successful, desirable women and men. Clambering to the top of the social hierarchy and finding a good mate was crucially important throughout human history, and we cannot now go back and erase that long heritage, just because it makes us queasy.

Body Shape

If physical attractiveness is the most important determinant of women's social dominance, then the most important determinant of physical attractiveness is body shape.

Women spend much more time thinking about the appearance of their body than men think about theirs, and they also report far more episodes of body comparison with others of their sex. When women are asked what bodily features they compare when looking at other women, they say they focus on other women's thighs, hips and waists, but are also interested in comparing body size, breasts, cleavage, cellulite and clothes. And when experimental subjects are shown pictures of women's bodies, they naturally assume that women with low waist–hip ratios are more attractive and socially dominant. Surveys confirm that popular and socially confident women are more likely to be near the societal ideal for body shape – and in the US and UK school, college and university environments in which many of these studies took place, that usually means being petite, slim and moderately curvy. Perhaps even more importantly, studies also indicate that girls and women *believe* that weight and shape are important for popularity and social success.

Girls and women who are dissatisfied with their bodies spend less time in social settings, have fewer friends, and worry they have less support from peers. They report that more of their social contacts with other women have a negative flavour to them, that they are less intimate with their friends and other women, and that they have less social influence. An extreme form of this occurs in women with eating disorders – who are usually perceived by themselves and others to be of low social rank, make more negative comparisons of their bodies with others, and measurably exhibit more submissive behaviours in social situations.

Interactions with other women are also crucial to the development of women's body image in the first place. For example, teenage girls say that their friends are especially important in establishing their own beliefs about weight and dieting. And indeed, members of female friendship groups are statistically more likely to have similar body mass indices, similar levels of body satisfaction, and similar attitudes to the importance of weight, dieting and body shape. There are of course many exceptions to this, but in general women with similar bodies and body-beliefs tend to end up being friends.

Social dominance has been with us for millions of years, but women's social worlds have now become intertwined with economics. Girls from families with higher socio-economic status are more likely to want a slimmer body shape, and tend to diet more. Higher socio-economic status women are on average thinner, exercise more, and dislike obesity more. These influences also get passed down the generations, because in many developed countries the children of such women worry more about becoming obese, and are half as likely to actually become obese as the children of lower-status women. For a woman, having a slim, small, feminine body correlates with her chances of acceptance by high socio-economic status groups, as well as her prospects for employment, income, housing, marriage and education – and furthermore, education itself correlates particularly strongly with the desire to be slim.

The links between body shape and socio-economic status are complex, but there must be forces at work in society which make higher-status people prefer female slimness in the first place. For example, sociologists have suggested that people in lower socio-economic groups may favour the 'functional' aspects of women's bodies, which allow them to work and provide for their families, whereas people in higher socio-economic groups are sufficiently removed from those everyday struggles to be able to focus more on the 'aesthetic' aspects of the female body. According to this argument, a petite, slim female body is a socio-economic symbol of being unencumbered by the demands of crude survival.

So body shape is the key element of physical appearance which dictates a woman's social dominance, but why? We have seen that body shape is something which men seek, and which indicates genetic and reproductive vigour, but the same is true of women's behaviour, facial beauty, and the thickness of their hair – yet none of those things is as important as body shape in the social rankings. I believe there are three interlinked reasons for this.

The first of these is that – as heterosexual men all seem to innately know – more than any other element of women's appearance, body shape is a reliable indicator of a woman's genetic health *and* her future ability to conceive and rear children. It is a visual distillation of her biological fortitude, and this, in short, is why it dictates her social status. Social dominance is not just an arbitrary game that human beings play – women reach the top of the pecking order for a reason, and that reason is that their bodies signal biological superiority.

A second feature of body shape is that today's superabundance of food means that women's bodies vary in shape far more than at any other time in the past. When humans were hunter-gatherers, or medieval serfs, or building the first corrals on the Great Plains, almost all women were slim. Yes, women's proportions – their 'ratios' – varied, and men used those ratios as sexual cues, but their absolute size differed little. Today, women's bodies are a diverse smorgasbord of curvaceousness from which men can choose, and within which women can compete. In contrast, facial attractiveness has become less varied in recent centuries as the disfiguring effects of disease, malnutrition and inbreeding have become a thing of the past. In other words, body shape has become more important than ever because it has become the most *variable* element of women's physical appearance.

The third reason is that women can actively alter their body shape in a profound way. Women can change their facial appearance by using cosmetics, and they can change their hair with a new cut, but these things are transient and unreal – they obscure, rather than reveal the woman beneath. We all know that when the makeup is washed off or the hairstyle grows out, the woman will be the same woman she was before. She achieved a temporary alteration to her superficial appearance, and may have looked more attractive as a result, but she cannot change her face or hair for ever. In contrast, women do have the ability to change their bodies. A woman can decide to gain or lose weight, and she can exercise to enhance her muscle tone – and everyone will know that those changes are real, medium-term (or longer) alterations to her body. Unlike every other aspect of visual appearance, body shape can truly be *changed*, offering the chance to change the way women feel about themselves and the way others view them.

Indefinableness

Yet we are left with an uneasy feeling. It seems somehow wrong, somehow unacceptable that in our feminism-inspired, diversity-valuing new society, men still use physical attraction as the prime drive to their choice of romantic, sexual and life-partners. We all enjoy falling in love with someone beautiful, but would find it hard to justify it as a politically correct decision. Our world is full of cultural complexity, dazzling technology and artistic wonder, yet we still select our romantic partners on the basis of whom we would like to wriggle around naked with for the next half-century, or whom we would like to see flickers of in the faces of our children.

All around us is evidence that a love of good looks is built into the human psyche. We tend to hook up with people of a similar level of attractiveness to ourselves, and 'similarly attractive' couples exhibit more public displays of affection. And of course we still vaunt 'love at first sight' as something to be sought out, and even envied. The logical part of our minds, even if it listens to people like me who argue this is all about natural selection and successful reproduction, still finds it unsettling that something as primal and unfair as bodily attractiveness continues to exert such control. This is probably why the following summary of this chapter seems disquieting:

$$\text{body shape} \rightarrow \text{visual attractiveness} \rightarrow \text{social dominance} \rightarrow \text{life success}$$

For men at least, the desire for beautiful female bodies is a consistent and uncomplicated process. Experiment after experiment has shown that visual bodily attractiveness is more important in men's choices of partners than in women's. Men also seem fairly consistent in their assessments of women's bodies – when shown images of different aspects of female bodies, they tend to agree on what they like, yet exhibit a flexibility to appreciate a wide range of female bodies which lie within their 'range of acceptability'. They are also judgemental about women's bodies, but as we have seen, their judgements may have very real validity – they assume, for example, that women with attractive bodies are healthier, more likely to be married, and more likely to have children. And finally, men are remarkably accurate – in one study they correctly predicted personality traits of women simply by viewing unposed photographs for as little as forty milliseconds.

Yet the fact remains that men's preferences for women seem to have little *direct* effect on how most women today want to look. Men's magazines and pornography are full of images of women with a wide range of body shapes, yet this male flexibility, this 'latitude' in male desire, does not seem to broaden society's idea of what constitutes an ideal female body shape. We all sense that there is a single body shape which is considered 'optimal', yet with this idea of a single optimum comes a paradox: the voices telling women how to appear can be strangely inconsistent. No one can tell us precisely, definitively, what that optimum actually is.

And this is where we finally encounter the thing that makes women's lives inherently more complicated than men's – a conflict in confidence that is a direct consequence of their body shape dictating their social standing. Because of its enormous biological importance, the shape of women's bodies has taken on a huge social importance too – and as a result each human society creates an ideal to which women are expected to aspire. This ideal usually falls somewhere within the wide range which men find attractive, yet it is itself far more narrowly defined than most men's own range of preferred body shapes. Societal ideals of female body shape are remarkably unforgiving, yet even worse, they are simultaneously frustratingly imprecise.

Women feel driven towards these ideals, yet no one tells them precisely what they are – what exactly they are meant to be seeking.

The media may tell women to be thin, but how thin is thin, and what happens when the media backtrack a little towards curviness? The ideal body shape is always changeable, malleable, ephemeral. It never exists as a clearly stated absolute which women can actually achieve – it remains forever out of reach, unattainable. Not only are women told to chase a moving target, but they are never even told where that moving target actually is. And of course they are then even criticised for being so shallow as to attempt to chase it.

The body-ideal is in continual flux. We have already seen that female body-ideals vary between societies, and that they change over time within individual cultures. Over the years fashions change, as the media openly fuel the obsession with women's bodies – representing them as talismans of the success or failure of individual women's lives. The publications which seek futilely to define the body-ideal career erratically between extremes, from the faux-life-affirming 'so-and-so shows off her new curves on the beach' to the weaselly 'Adele – every inch a star!' Even women's own internal biology conspires against them, as their own body shape and body image change cyclically over the course of every month. High levels of oestrogen around the time of ovulation reduce women's assessments of other women's attractive-ness, and they also reduce their appetite for food, probably more so if they are high up in the dominance hierarchy. Body-ideals change all the time, as do the body-self-images that women are continually told to compare with them. Women never know where they stand.

In contrast, men usually have a clear idea of their place in the pecking order. Men's dominance interactions often involve shows of strength, sporting prowess or wealth, which are meticulously scored or measured to generate clear, unchallengeable outcomes. Failing that, they resort to fighting, in which one combatant usually wins an obvious victory. Even at school, little boys may slug it out from time to time, but these encounters are often decisive and rapidly lead to the establishment of a peaceful playground entente. And in later life, male social hierarchies usually remain relatively clear-cut, certain and accepted by all, leading to a more or less frictionless acceptance of the status quo – largely because men do not establish their dominance hierarchy on the basis of something as vague as bodily beauty.

And it is this vagueness, this imprecision of the main criterion of female social rank which makes life inescapably complex for women. Men demonstrate their status by *doing* something (simple), women by *being* something (complex). Social rank is immensely important to all humans' health and happiness, but it is the unavoidable uncertainty in women's dominance hierarchies which explains why their relationships with their own bodies must always be much more complex than men's.

This is why women think about their bodies so much – why, so much more than men, they become locked in repeated cycles of analysis, assessment and judgement of their own bodies, to a level of detail and complexity which men find inexplicable and bewildering. A woman's body is the most important part of the game, but someone keeps changing the rules.

Escape

The drive to gain social dominance by having a beautiful body is so powerful that it often trumps everything else – the health, physical prowess, intellectual and economic achievement which beauty and dominance are actually meant to signify. It also trumps the very objective for which dominance exists – the need to be attractive to the opposite sex. The many powers of the female body have been reduced to this one imperative – to have a body which impresses yourself and impresses other women. But why do human females behave in such an apparently perverse way, in which the social means overpower the biological ends? What is it about humans which allows this state of affairs to persist? As a species, we have succeeded mainly because we have enormous brains which construct elaborate patterns of thought, but maybe this is one case in which the brain has 'over-thought' the task in hand. Social dominance is beneficial because it serves a purpose, but only a huge, unstable brain like ours could take women to the point where the drive to be dominant becomes so all-consuming, so self-defeating.

Maybe human women once coped with this paradox reasonably happily, so long as their social world was relatively simple. We believe that for most of human history, humans lived in small, multi-family groups of no more than two hundred people. In many parts of the world our ancestors may have been so sparsely distributed that each of these groups may only rarely have glimpsed another human tribe on the horizon and women had endless days to develop a coherent social interaction with the few women around them. They did not have to apply their cosmetic war-paint every Friday to impress yet another cohort of previously unencountered women at the bar or club. They never saw pouting images of women they did not know with bodies they did not recognise. They never had to withstand the appearance-based snap decisions of internet dating. Everyone's social world was circumscribed by the limits of the tribe, and maybe this provided a stability, a long-lost certainty, to the female social hierarchy. Indeed, my interviewee 'C' even said to me: 'Perhaps that's why I go to the gym so much – those people, they're my *tribe*.'

But do women *have* to be locked in a prison of self-surveillance, enchained by the idea that they must view their bodies as others view them if they are to self-criticise and self-improve successfully? And, whose fault is this?

It has been argued that today it is women, not men, who perpetuate this system which consigns women to a life of subordinate visual self-vigilance. For much of the human story, it was probably men, the patriarchy, who were to blame, but men are no longer solely culpable. As we have seen, women criticise other women's bodies much more often than men do and women have less forgiving ideas about what constitutes an attractive female body. And throughout the animal kingdom dominant females suppress subordinate females' sexuality to make sex a more valuable bargaining chip with which to extract favours from males. Some feminist writers have claimed that it is now women who most clearly articulate society's body-chauvinism – that having set the ball rolling, men are now standing back and letting women become the enforcers, their own worst enemies.

Today, female body shape is an inescapable, exploitable commodity – and many women feel torn between the body they live in and the body they must advertise to the world. Yet we also stand at a turning-point in our species' history where we possess enough self-awareness to decide if we want to challenge the instinctive hold which female body shape exerts over us. Over the last few million years, humans have acquired an exceptional body, and women have acquired an especially exceptional variant of it – with a unique set of curves, and the fertility, vitality, appetites, thoughts and fears which go with them. Those curves have become the key driver behind women's social status. They are signifiers of supposed 'quality' which we humans cannot at present ignore, no matter how the idea repels us.

But can we escape our obsession with the female body? Do we even want to? After all, the prime, original difference between the sexes is that their bodies are different. One gestates and one does not, and that is not going to change any time soon. The physical, evolutionary, psychological and social differences between women and men are founded upon that obvious and all-important difference: this is the root of heterosexual attraction and the impulse on which our species depends. But will we forever find it acceptable that men love their partners partly because of this base desire for their bodies? Can we, by a conscious decision, change the criteria by which our social hierarchies are established? Can we ever create a world in which our daughters would rather be called intelligent or successful, than beautiful?

Increasingly, everywhere, there is variety – some women challenge the prevailing ideal, some are homosexual, some choose partners from cultures with different body ideals. Our species is changing at an unprecedented rate, and we do not yet know if we will wish to drag our ancient urges and chauvinisms along with us on our journey. The ages-old power of female body shape is strong, it lies deep within us all, and taming its power will be the ultimate test of human self-determination.

Acknowledgements

I would like to thank the Penguin Group for permission to use the quotation from *Stranger in a Strange Land* at the start of the introduction. Also, I would like to thank John Irving and the Cooke Agency for their permission to use the quotation from *Last Night in Twisted River* at the start of chapter one.

In writing this book, it has been my pleasure to discuss its slowly gestating contents with some real experts. First of all, my thanks go to Natasha Devon of bodygossip.org for bumping into me at the time when I was first planning the structure of this book. Professor Ali Mobasheri was extremely helpful in guiding me towards useful information about the effect of female body shape on disease, and Dr Nick Medford gave me lots of good tips about the science of the sense of inhabiting the body. Also, I would like to thank my five wonderful interviewees for putting up with my inept questioning. Finally, and as always, I would like to thank my agent Peter Tallack at the Science Factory for his continued help and support, and my editor at Portobello, Laura Barber, for her enthusiasm and putting up with my idiosyncratic text.

Selected Bibliography

1. Where women's bodies came from

Aiello, L. and Dean, C. (1990). *An Introduction to Human Evolutionary Anatomy*. London: Academic Press.

Dahlnerg, F. (1981). *Woman the Gatherer*. New Haven: Yale University Press.

Kurki, H.K. (2013). Bony pelvic canal size and shape in relation to body proportionality in humans. *American Journal of Physical Anthropology* 151, 88–101.

Lassek, W.D. and Gaulin, S.J.C. (2008). Waist–hip ratio and cognitive ability: is gluteofemoral fat a privileged store of neuro-developmental resources? *Evolution and Human Behaviour* 29, 26–34.

LaVelle, M. (1995). Natural selection and developmental sexual variation in the human pelvis. *American Journal of Physical Anthropology* 98, 59–72.

Lovejoy, C.O. (2005). The natural history of human gait and posture Part 1. Spine and pelvis. *Gait and Posture* 21, 95–112.

Morgan, M.H. and Carrier, D.R. (2013). Protective buttressing of the human fist and the evolution of hominin hands. *Journal of Experimental Biology* 216, 236–44.

Penin, X., Berge, C. and Baylac, M. (2002). Ontogenetic study of the skull in modern humans and the common chimpanzees: neotenic hypothesis reconsidered with a tridimensional Procrustes analysis. *American Journal of Physical Anthropology* 118, 50–62.

Stewart D.B. (1984). The pelvis as a passageway. I. Evolution and adaptations. *British Journal of Obstetrics and Gynaecology* 91, 611–17.

Wells, J.C.K. (2010). *The Evolutionary Biology of Human Fatness.* Cambridge: Cambridge University Press.

Weston, E.M., Friday, A.E. and Liò, P. (2007). Biometric evidence that sexual selection has shaped the hominin face. *PLoS ONE* 2, e710.

Whitcome, K.K., Shapiro, L.J. and Lieberman, D.E. (2007). Fetal load and the evolution of lumbar lordosis in bipedal hominins. *Nature* 450, 1075–8.

2. Where women's bodies come from

Bainbridge, D.R.J. (2003). *The X in Sex: How the X Chromosome Controls our Lives.* Cambridge, MA: Harvard University Press.

Cash, T.F. (2012). *Encyclopaedia of Body Image and Human Appearance.* London: Academic Press.

Dufour, D.L. and Sauther, M.L. (2002). Comparative and evolutionary dimensions of the energetics of human pregnancy and lactation. *American Journal of Human Biology* 14, 584–602.

Khan, M.H., Victor, F., Rao, B. and Sadick, N.S. (2010). Treatment of cellulite: Part I. Pathophysiology. *Journal of the American Academy of Dermatology* 62, 361–70.

Lassek, W.D. and Gaulin, S.J. (2006). Changes in body fat distribution in relation to parity in American women: a covert form of maternal depletion. *American Journal of Physical Anthropology* 131, 295–302.

Pond, C.M. (1998). *The Fats of Life.* Cambridge: Cambridge University Press.

Rebuffé-Scrive, M. et al. (1985). Fat cell metabolism in different regions in women. Effect of menstrual cycle, pregnancy, and lactation. *The Journal of Clinical Investigation* 75, 1973–6.

Rosenbaum, M. and Leibel, R.L. (1999). Role of gonadal steroids in the sexual dimorphisms in body composition and circulating concentrations of leptin. *Journal of Clinical Endocrinology and Metabolism* 84, 1784–9.

Stevens, J., Katz, E.G. and Huxley, R.R. (2010). Associations between gender, age and waist circumference. *European Journal of Clinical Nutrition* 64, 6–15.

Weisfeld, G.E. (1999). *Evolutionary Principles of Human Adolescence.* New York: Basic Books.

Zafon, C. (2007). Oscillations in total body fat content through life: an evolutionary perspective. *Obesity Reviews* 8, 525–30.

3. The power of curves

Bainbridge, D.R.J. (2009). *Teenagers: A Natural History.* London: Portobello.

Ellison, P.T. (2003). Energetics and Reproductive Effort. *American Journal of Human Biology* 15, 342-51.

Eriksson, N. and others (2010). Genetic variants associated with breast size also influence breast cancer risk. *BMC Medical Genetics* 13, 53.

Evans, D.J., Hoffmann, R.G, Kalkhoff, R.K. and Kissebah, A.H. (1984). Relationship of body fat topography to insulin sensitivity and metabolic profiles in premenopausal women. *Metabolism* 33, 68-75.

Fagherazzi, G. et al. (2012). Hip circumference is associated with the risk of premenopausal ER-/PR- breast cancer. *International Journal of Obesity* 36, 431–9.

Frisch, R.E. (2002). *Female Fertility and the Body Fat Connection.* Chicago: University of Chicago Press.

Kaplowitz, P.B. (2008). Link between body fat and the timing of puberty. *Pediatrics* 121, S208.

Lassek, W.D. and Gaulin, S.J. (2007). Menarche is related to fat distribution. *American Journal of Physical Anthropology* 133, 1147–51.

Norgan, N.G. (1997). The beneficial effects of body fat and adipose tissue in humans. *International Journal of Obesity and Related Metabolic Disorders* 21, 738–46.

Radzevičienè, L. and Ostrauskas, R. (2013). Body mass index, waist circumference, waist–hip ratio, waist–height ratio and risk for type 2 diabetes in women: a case-control study. *Public Health* 127, 241–6.

Shay, C.M. et al. (2010). Regional adiposity and risk for coronary artery disease in type 1 diabetes: does having greater amounts of gluteal-femoral adiposity lower the risk? *Diabetes Research and Clinical Practice* 89, 288–95.

4. What men want and why it doesn't matter

Barelds-Dijkstra, P. and Barelds, D.P. (2008). Positive illusions about one's partner's physical attractiveness. *Body Image* 5, 99–108.

Bereczkei, T., Gyuris, P. and Weisfeld, G.E. (2004). Sexual imprinting in human mate choice. *Proceedings of the Royal Society: Biological Sciences* 271, 1129–34.

Cellerino, A. (2003). Psychobiology of facial attractiveness. *Journal of Endocrinological Investigations* 26, supplement 3, 45–8.

Crossley, K.L., Cornelissen, P.L. and Tovée, M.J. (2012). What is an attractive body? Using an interactive 3D program to create the ideal body for you and your partner. *PLoS One* 7, e50601.

Dixson, B.J., Grimshaw, G.M., Linklater, W.L. and Dixson, A.F. (2011). Eye-tracking of men's preferences for waist-to-hip ratio and breast size of women. *Archives of Sexual Behaviour* 40, 43–50.

Garver-Apgar, C.E., Gangestad, S.W., Thornhill, R., Miller, R.D. and Olp, J.J. (2006). Major histocompatibility complex alleles, sexual responsivity, and unfaithfulness in romantic couples. *Psychological Science* 17, 830–5.

Hayes, A.F. (1995). Age preferences for same- and opposite-sex partners. *Journal of Social Psychology* 135, 125–33.

Holliday, I.E., Longe, O.A., Thai, N.J., Hancock, P.J. and Tovée, M.J. (2011). BMI not WHR modulates BOLD fMRI responses in a subcortical reward network when participants judge the attractiveness of human female bodies. *PLoS One* 6, e27255.

Kanazawa, S. (2011). Intelligence and physical attractiveness. *Intelligence* 39, 7–14.

LeBlanc, S.A. and Barnes, E. (1974) . On the adaptive significance of the female breast. *American Naturalist* 108, 577–8.

Miller, G. (2000). Sexual selection for indicators of intelligence.

Novartis Foundation Symposia 233, 260–70.

Møller, A.P. (1995). Breast asymmetry, sexual selection, and human reproductive success. *Ethology and Sociobiology* 16, 207–19.

Platek, S.M. and Singh, D. (2010). Optimal waist-to-hip ratios in women activate neural reward centers in men. *PLoS One* 5, e9042.

Sefcek, J.A., Brumbach, B.H., Vasquez, G. and Miller, G.F. (2006). The evolutionary psychology of human mate choice: how ecology, genes, fertility and fashion influence mating strategies. In Kauth, M.R. ed. *Handbook of the Evolution of Human Sexuality*. New York: Haworth Press.

Singh, D., Dixson, B.J., Jessop, T.S, Morgan, B. and Dixson, A.F. (2010) Cross-cultural consensus for waist–hip ratio and women's attractiveness. *Evolution and Human Behaviour* 31, 176–81.

Voracek, M. and Fisher, M.L. (2006). Success is all in the measures: androgenousness, curvaceousness, and starring frequencies in adult media actresses. *Archives of Sexual Behaviour* 35, 297–304.

5. Trapped in a vessel of flesh

Bainbridge, D.R.J. (2012). *Middle Age: A Natural History*. London: Portobello.

Berlucchi, G. and Aglioti, S.M. (2009). The body in the brain revisited. *Experimental Brain Research* 200, 25–35.

Boyes, A.D. and Latner, J.D. (2009). Weight stigma in existing romantic relationships. *Journal of Sex and Marital Therapy* 35, 282–93.

Cash, T.F. and Smolak, L. (2011). *Body Image*. New York: Guilford Press.

Ehrsson, H.H., Kito, T., Sadato, N., Passingham, R.E. and Naito, E. (2005). Neural substrate of body size: illusory feeling of shrinking of the waist. *PLoS Biology* 3, e412.

Hodzic, A., Muckli, L., Singer, W. and Stirn, A. (2009). Cortical responses to self and others. *Human Brain Mapping* 30, 951–61.

Kurosaki, M., Shirao, N., Yamashita, H., Okamoto, Y. and Yamawaki, S. (2006). Distorted images of one's own body activates the prefrontal cortex and limbic/paralimbic system in young women:

a functional magnetic resonance imaging study. *Progress in Neuro-psychopharmacology and Biological Psychiatry* 59, 380–6.

Pfeifer, R. and Bongard, J. (2007) *How the Body Shapes the Way we Think.* Cambridge, MA: MIT Press.

Saxe, R., Jamal, N. and Powell, L. (2006). My body or yours? The effect of visual perspective on cortical body representations. *Cerebral Cortex* 16, 178–82.

Smolak, L. (2004). Body image in children and adolescents: where do we go from here? *Body Image* 1, 15–28.

6. Comfort and discomfort eating

Bautista, C.J., Martínez-Samayoa, P.M. and Zambrano, E. (2012). Sex steroids regulation of appetitive behavior. *Mini Reviews in Medical Chemistry* 12, 1107–18.

Dulloo, A.G., Jacquet, J. and Montani, J.P. (2012). How dieting makes some fatter: from a perspective of human body composition autoregulation. *Proceedings of the Nutrition Society* 71, 379–89.

Grogan, S. (2008) *Body Image.* Hove: Routledge.

Gruber, A.J., Pope, H.G. Jr, Lalonde, J.K. and Hudson, J.I. (2001). Why do young women diet? The roles of body fat, body perception, and body ideal. *Journal of Clinical Psychiatry* 62, 609–11.

Kemps, E. and Tiggemann, M. (2005). Working memory performance and preoccupying thoughts in female dieters: evidence for a selective central executive impairment. *British Journal of Clinical Psychology* 44, 357–66.

Markus, C.R. et al. (1998). Does carbohydrate-rich, protein-poor food prevent a deterioration of mood and cognitive performance of stress-prone subjects when subjected to a stressful task?. *Appetite* 31, 49–65.

Murray, S. (2008). *The Fat Female Body.* Basingstoke: Palgrave Macmillan.

Prichard, I. and Tiggemann, M. (2008). An examination of pre-wedding body image concerns in brides and bridesmaids. *Body Image* 5, 395–8.

Ogden, J. (2010). *The Psychology of Eating*. Chichester: Wiley-Blackwell.

Raichlen, D.A., Foster, A.D., Gerdeman, G.L., Seillier, A. and Giuffrida, A. (2012). Wired to run: exercise-induced endocannabinoid signaling in humans and cursorial mammals with implications for the 'runner's high'. *Journal of Experimental Biology* 215, 1331–6.

Rolls, B.J., Fedoroff, I.C. and Guthrie, J.F. (1991). Gender differences in eating behavior and body weight regulation. *Health Psychology* 10, 133–42.

7. A malaise of shapes

Abed, R.T. (1998). The sexual competition hypothesis for eating disorders. *British Journal of Medical Psychology* 71, 525–47.

Castellini, G. et al. (2013). Looking at my body. Similarities and differences between anorexia nervosa patients and controls in body image visual processing. *European Psychiatry* 28, 427–35.

Chamay-Weber, C., Narring., F. and Michaud, P.A. (2005). Partial eating disorders among adolescents: a review. *Journal of Adolescent Health* 37, 417–27.

Fessler, D.M. (2002). Pseudoparadoxical impulsivity in restrictive anorexia nervosa: a consequence of the logic of scarcity. *International Journal of Eating Disorders* 31, 376–88.

Gatward, N. (2007). Anorexia nervosa: an evolutionary puzzle. *European Eating Disorders Review* 15, 1–12.

Gaudio, S. and Quattrocchi, C.C. (2012). Neural basis of a multi-dimensional model of body image distortion in anorexia nervosa. *Neuroscience and Biobehavioral Reviews* 36, 1839–47.

Guisinger, S. (2003). Adapted to flee famine: adding an evolution-ary perspective on anorexia nervosa. *Psychological Review* 110, 745–61.

Jansen, A., Nederkoorn, C. and Mulkens, S. (2005). Selective visual attention for ugly and beautiful body parts in eating disorders. *Behavioural Research and Therapy* 43, 183–96.

Kaye, W. (2008). Neurobiology of anorexia and bulimia nervosa. *Physiology and Behavior* 94, 121–35.

Klump, K.L., Keel, P.K., Sisk, C. and Burt, S.A. (2010). Preliminary evidence that estradiol moderates genetic influences on disordered eating attitudes and behaviors during puberty. *Psychological Medicine* 40, 1745–53.

Klump, K.L., Suisman, J.L., Burt, S.A., McGue, M. and Iacono, W.G. (2009). Genetic and environmental influences on disordered eating: An adoption study. *Abnormal Psychology* 118, 797–805.

Morgan, H.G. (1977). Fasting girls and our attitudes to them. *British Medical Journal* 2, 1652–5.

Nasser, M., Baistow, K. and Treasure, J. (2007). *The Female Body in Mind*. London: Routledge.

8. Following the fashion

Allison, D.B., Hoy, M.K., Fournier, A. and Heymsfield, S.B. (1993). Can ethnic differences in men's preferences for women's body shapes contribute to ethnic differences in female adiposity? *Obesity Research* 1, 425–32.

Bair, C.E., Kelly, N.R., Serdar, K.L. and Mazzeo, S.E. (2012). Does the Internet function like magazines? An exploration of image-focused media, eating pathology, and body dissatisfaction. *Eating Behaviors* 13, 398–401.

Byrd-Bredbenner, C., Murray, J. and Schlussel, Y.R. (2005). Temporal changes in anthropometric measurements of idealized females and young women in general. *Women's Health* 41, 13–30.

Calado, M., Lameiras, M., Sepulveda, A.R., Rodríguez, Y. and Carrera, M.V. (2010). The mass media exposure and disordered eating behaviours in Spanish secondary students. *European Eating Disorders Review* 18, 417–27.

Dittmar, H., Halliwell, E. and Ive, S. (2006). Does Barbie make girls want to be thin? The effect of experimental exposure to images of dolls on the body image of 5- to 8-year-old girls. *Developmental Psychology* 42, 283–92.

Field, A.E., Cheung, L., Wolf, A.M., Herzog, D.B., Gortmaker, S.L. and Colditz, G.A. (1999). Exposure to the mass media and weight concerns among girls. *Pediatrics* 103, E36.

Friederich, H.C. et al. (2007). I'm not as slim as that girl: neural bases of body shape self-comparison to media images. *Neuroimage* 15, 674–81.

Mazzeo, S.E., Trace, S.E., Mitchell, K.S. and Gow, R.W. (2007). Effects of a reality TV cosmetic surgery makeover program on eating disordered attitudes and behaviors. *Eating Behaviors* 8, 290–7.

Suisman, J.L. et al. (2012). Genetic and environmental influences on thin-ideal internalization. *International Journal of Eating Disorders* 45, 942–8.

Swami, V. et al. (2010). The attractive female body weight and female body dissatisfaction in 26 countries across 10 world regions: results of the international body project I. *Personality and Social Psychology Bulletin* 36, 309–25.

9. Covering up and tucking in

Boyce, J.A., Martens, A., Schimel, J. and Kuijer, R.G. (2012). Preliminary support for links between media body ideal insecurity and women's shoe and handbag purchases. *Body Image* 9, 413–16.

Cash, T.F., Dawson, K., Davis, P., Bowen, M. and Galumbeck, C. (1989). Effects of Cosmetics Use on the Physical Attractiveness and Body Image of American College Women. *Journal of Social Psychology* 129, 349–55.

Elliot, A.J. and Niesta, D. (2008). Romantic red: red enhances men's attraction to women. *Journal of Personality and Social Psychology* 95, 1150–64.

Etcoff, N.L., Stock, S., Haley, L.E., Vickery, S.A. and House, D.M. (2011). Cosmetics as a feature of the extended human phenotype: modulation of the perception of biologically important facial signals. *PLoS One* 6, e25656.

Fielding, R. et al. (2008). Are longer legs associated with enhanced fertility in Chinese women? *Evolution and Human Behaviour* 29, 434–43.

Flensmark, J. (2004). Is there an association between the use of heeled footwear and schizophrenia? *Medical Hypotheses* 63, 740–7.

Grammer, K., Renninger, L. and Fischer, B. (2004). Disco clothing, female sexual motivation, and relationship status: is she dressed to impress? *Journal of Sex Research* 41, 66–74.

Guéguen, N. (2013). Effects of a Tattoo on Men's Behavior and Attitudes Towards Women: An Experimental Field Study. *Archives of Sexual Behavior* 42, 1517–24.

Haselton, M.G., Mortezaie, M., Pillsworth, E.G., Bleske-Rechek, A. and Frederick, D.A. (2007). Ovulatory shifts in human female ornamentation: near ovulation, women dress to impress. *Hormones and Behavior* 51, 40–5.

Thompson, P. and Mikellidou, K. (2011). Applying the Helmholtz illusion to fashion: horizontal stripes won't make you look fatter. *i-Perception* 2, 69–76.

Webb, W.M. (1908). *The Heritage of Dress Being Notes on the History, and Evolution of Clothes.* New York: McClure.

10. Why women care and why it's complicated

Arciszewski, T., Berjot, S. and Finez, L. (2012). Threat of the thin-ideal body image and body malleability beliefs: effects on body image self-discrepancies and behavioral intentions. *Body Image* 9, 334–41.

Baumeister, R.F. and Twenge, J.F. (2002). Cultural Suppression of Female Sexuality. *Review of General Psychology* 6, 166–203.

Benzeval, M., Green, M.J. and Macintyre, S. (2013). Does perceived physical attractiveness in adolescence predict better socioeconomic position in adulthood? Evidence from 20 years of follow up in a population cohort study. *PLoS One* 22, e63975.

Campbell, A. (1999). Staying alive: evolution, culture, and women's intrasexual aggression. *The Behavioral and Brain Sciences* 22, 203–14.

Cashdan, E (1998). Are men more competitive than women? *British Journal of Social Psychology* 37, 213–29.

Eckert, P. (1990). Cooperative competition in adolescent 'girl talk'. *Discourse Processes* 13, 91–122.

Fisher, M.L. (2004). Female intrasexual competition decreases female facial attractiveness. *Proceedings of the Royal Society of London Series B Supplement* 271, S283–S285.

Försterling, F., Preikschas, S. and Agthe, M. (2007). Ability, luck, and looks: an evolutionary look at achievement ascriptions and the sexual attribution bias. *Journal of Personality and Social Psychology* 92, 775–88.

Gilbert, P. (2002). Relationship of anhedonia and anxiety to social rank, defeat and entrapment. *Journal of Affective Disorders* 71, 141–51.

Jokela, M. (2009). Physical attractiveness and reproductive success in humans: evidence from the late 20th century United States. *Evolution and Human Behavior* 30, 342–50.

Kennedy, J.H. (1990). Determinants of peer social status: contributions of physical appearance, reputation, and behaviour. *Journal of Youth and Adolescence* 19, 233–44.

Levin, J. and Arluke, A. (1985). An Exploratory Analysis of Sex Differences in Gossip 1. *Sex Roles* 12, 281–6.

Michopoulos, V. and Wilson, M.E. (2011). Body weight decreases induced by estradiol in female rhesus monkeys are dependent upon social status. *Physiology and Behavior* 102, 382–8.

Stockley, P. and Bro-Jørgensen, J. (2011). Female competition and its evolutionary consequences in mammals. *Biological Reviews of the Cambridge Philosophical Society* 86, 341–66.

Townsend, J.M. and Wasserman, T. (1997). The perception of sexual attractiveness: sex differences in variability. *Archives of Sexual Behavior* 26, 243–68.

Troop, N.A. and Baker, A.H. (2008). The specificity of social rank in eating disorder versus depressive symptoms. *Eating Disorders* 16, 331–41.

Index

abdomens: abdominal fat and health, 52; breastfeeding's effect on size, 39; and clothing, 173–4; differences between the sexes, 17, 36; societal ideal through the ages, 147, 148–9
abdominoplasty, 186
Abercrombie and Fitch, 187
Adam, 164
adipocytes, 29–31, 35
African ideal female body shape, 153–4, 155
African-American ideal female body shape, 151–2
age: and body-dissatisfaction, 95–8; and eating disorders, 127; and sexual selection, 69–70, 79
aggression, 193–5
Agta people, 25
amphetamine-regulated transcript, 110
amygdala, 131, 132–3
Andaman Islanders, 155
androgens, 36–7, 73
anger: and body image, 101; and weight gain, 113
anorexia nervosa, 123–41; non-fat-phobic, 153; *see also* eating disorders
antidepressants, and eating disorders, 133
anxiety: and appetite, 113; and eating disorders, 129
appetite, 30, 108–15, 119–20
Aqua, 145–6
areolae, 79

arms, 12, 14, 16–17, 35, 40, 174
aroma *see* smell
art, and sexual selection, 67
Artemis statue, Ephesus, 147
arthritis, 54
Asian ideal female body shape, 153
athletic abilities, 24–6
atherosclerosis, 54
attractiveness: children's attitude to, 85, 95; and goodness, 102–3, 121–2, 196; and health, 65–6, 197–8; and social status, 195–200; *see also* sexual attraction and selection
Australopithecus, 19–20, 24
autism, and eating disorders, 129

backs and spines, 12, 20, 40
Baghavad-Gita, 61
Bailey, David, 167
ballet dancers, 120
Bantu people, 155
Barbie, 145–6, 160–1
Bardot, Brigitte, 82
beauty contests, 153–4, 187
beer, 34
bellies *see* abdomens
Bible, 104
bipedalism, 11–12
black ideal female body shape, 151–2
blood pressure, 53
body-dissatisfaction, 95–8, 120
Body Dysmorphic Disorder, 124
body image, 84–103; avoiding, 100; checking, 99; coping with, 99; and eating disorders, 130–2; ethnosocial

variations, 151–6; fixing, 99; and the media, 153, 156–63; and social status, 199–200

body mass index: calculating, 8, 46–7; and health, 52; and sexual selection, 73–4, 80

body-perception, 93–5

body-relevance, 97–8

body shape, female: ability to be changed, 201–2; body-model concept, 92; changes throughout life, 27–41; differences from male, 13–28, 31–44; and eating disorders, 134–6; escaping the obsession, 205–7; and evolution, 9–26, 42–4; female disregard for male preferences, 3, 82–3; female obsession with, 2–3, 85–6, 189–205; female perception of, 84–103; female preferences, 80–1; and health, 201; and inheritance, 42; male preferences, 61–83, 152, 156, 175–6, 183, 202–3; and social status, 199–201; societal ideal and the media, 153, 156–63; societal ideal and its pressures, 146, 156–63; societal ideal in different ethnosocial groups, 155–6; societal ideal in flux, 203–5; societal ideal through the ages, 146–51; symmetry, 71; teasing about, 134; uniqueness of human, 2, 5; variations in female, 41–4, 65, 81–2; ways of altering and enhancing, 164–88

body-tending activities, 99–100

bones, 17, 54

boredom, and appetite, 113

botox, 186–7

bottoms see buttocks

brain: brain regions which identify bodies, 88–93; brain size and body shape, 15, 18–21; brain size's relation to mother's size, 14; cardiovascular disease, 53–4, 175; development pre- and post-birth, 18–19; dieting and mood, 119; eating and mood, 108–15; and eating disorders, 129–33, 141; energy consumption, 37; evolution, 18; exercise and mood, 120; and female fat stores, 23; and fertility, 49–50; relationship to the body, 86–93, 99–103; and sexual selection, 68–9; size dimorphism, 16; and walking upright, 12; and wearing high-heeled shoes, 177

bras, 171, 172

breastfeeding see lactation and breastfeeding

breasts: breast size and health, 54–7; cancer, 54, 55–7; and clothing, 170–2; cosmetic surgery, 185–6; development, 33; and fat deposition, 21, 22; female dissatisfaction with own, 97; lactation's effect on, 39; large breasts and promiscuity, 158; post-menopause, 40; reason for size of human breasts, 77–80; and sexual selection, 76–81; societal ideal through the ages, 147, 149, 150; waist–bust ratio, 76–7, 80–1

bulimia, 123–41; see also eating disorders

Burkina Faso, 154

Burma, 173

buttocks: and breastfeeding, 23, 39; and clothing, 170–3; cosmetic surgery, 186; development, 33; and fat deposition, 21, 22, 23, 31, 35; and pregnancy, 39; prominence, 12–13; and sexual selection, 63–4; societal ideal through the ages, 147, 148–9; steatopygia, 155

cancer, 21, 35, 54, 55–7, 175

carbohydrate, 110–11

cardiovascular disease, 53–4, 175

carrying angle see elbow angulation

cellulite, 36

cheeks, 36, 72, 183

chests, 17

chewing, 15, 109

childbirth: and being overweight, 48; and female size, 14; and pelvis size, 18–21; and the seasons, 22; *see also* pregnancy

childcare: feeding babies, 23, 37–9; and female athletic abilities, 25–6; and female fat deposition, 22; and the hips, 20–1

children: and body image, 95–6; and clothes, 169; differences between the sexes, 31–4; eating habits, 112–13; and embodiment, 85–6; overweight children and eating disorders, 134; reactions to fatness, 85, 95–7

chimpanzees: birth, 19; faces, 16; female competition among, 191; sexual selection, 69; skeleton shape, 13; weight dimorphism, 14

China, 187

chins, 72

chocolate, 114–15

chromosomes, 28–9

cinema, and ideal female body shape, 157–8

cingulate gyrus, 132–3

class *see* social status

cleavage, 171

clitorises, 37

clothing, 164–82; to alter appearance, 168–82; colour, 178–81; as concealment, 166–7; and gender-specificity, 168, 178–9; and menstrual cycle, 180, 198; origins, 166; as protection, 167–8; and sexual arousal, 167, 168; stripes, 181–2

cocaine, and dieting, 118

cocaine-regulated transcript, 110

colour: of clothing, 178–81; mammals' discernment of, 179

colour blindness, 86

competition, 190–8, 203–6; historical form, 206

contraception, and body image, 101

corsets, 149, 171

cortisol, 113

cosmetic surgery, 185–7

cosmetics *see* makeup

Cotard's syndrome, 93

Croft, Lara, 161

Cycladic civilisation and figures, 148, *148*

Darwin, Charles, 63

dementia, 54

depersonalisation disorder, 93

depilation, 184

depression: and chocolate, 115; and dieting, 116, 118, 119

Descartes, René, 84

diabetes, 51–4, 175

diet *see* food and eating

dieting, 105, 116–19, 134

disruptive selection, 43

division of labour, 24–6, 190

dogs, 120

dolls, and ideal female body shape, 145–6, 160–1

dominance: historical form, 206; importance to women, 190–8; uncertain nature of criteria for women, 203–5

dopamine, 133

eating *see* food and eating

eating disorders, 123–41; among African-Americans, 151; and brain function, 129–33, 141; definition, 123–6; and evolution, 136–40; and exercise, 120; among Hispanics, 152–3; historical incidences, 136; male, 126–7, 136; the media's influence, 160; partial, 126; resistance to treatment, 140–1; and social status, 199; sociocultural explanations, 133–6; websites, 161

elbow angulation, 16–17

embodiment, 84–103

emotions: and eating disorders, 132–3; *see also* mood

endocannabinoids, 110, 115, 120

endogenous opioids, 110
energy availability, and fertility, 49–50
enteroception, 91
Eve, 104, 149, 164
evolution: and body shape, 9–26,
    42–4; disruptive selection, 43; and
    eating disorders, 136–40; by natural
    selection, 63; by sexual selection,
    63–83
exercise: differences between the sexes,
    36; effects on women, 119–21;
    exercise dependence, 126; and
    fertility, 46, 48; among Sudanese
    women, 153; *see also* athletic
    abilities
eyebrows, 72
eyes, 12, 71, 72, 73

faces: decrease in variation, 201;
    development of facial recognition
    abilities, 85; differences between the
    sexes, 15–16; and sexual selection,
    69, 70–3
families *see* children; parents; siblings
fashion, 169
fashion models, 82, 177–8
fat: abdominal, 52–3; changes in
    body stores throughout life, 31–41;
    different types in female body,
    30–1; and fertility, 45–51; function,
    21–2; and health, 52–7; and division
    of labour, 24–6; liposuction, 186;
    measuring gluteofemoral, 48–9;
    and reproduction, 22–4; science
    behind function, 29–30; and sexual
    selection, 73–6; and variations in
    female body shape, 44
fat-talk, 96
feet, 12, 16, 88, 174, 176
femininity, and sexual selection, 65–6,
    67, 72–3
Feral Cheryl, 145
fertility: and body shape, 45–51;
    clothing as guide to, 180–1; and
    competition, 191; and eating
    disorders, 139–40; and leg length,

175; and menstrual cycle, 180; and
    sexual selection, 65–6
figure skating, 120
fingers, 16, 125
Florence, Brancacci Chapel, 164
food and eating: comfort foods,
    110–11; cravings, 113, 114–15;
    diet and fertility, 49; dieting, 105,
    116–19, 134; food availability and
    ideal female body shape, 155; food's
    relationship with mood, 108–15;
    guilt about, 111–12; instinctive urge
    to eat, 105; male and female eating
    habits, 108; *see also* eating disorders
fractures, 17

gall bladder, 54
genetics *see* inheritance and genetics
ghrelin, 110, 113
gorillas, faces, 16
gossip, 194
gynaecomastia, 54–5

hair: blonde, 183; depilation, 184;
    pubic and armpit, 36–7, 166, 184;
    and sexual selection, 70
hands, 16, 86–7, 174, 176–7
happiness, and appetite, 113
Harry, Debbie, 82
health: and attractiveness, 65–6, 197–8;
    and body shape, 201; and body
    symmetry, 71; and dieting, 118; and
    intelligence, 67; and leg length, 175;
    and skin tone, 73; and weight, 51–7
heart, 24, 53–4, 125
height, 36
Heinlein, Robert, 1
Helmholz illusion, 182
Hepburn, Audrey, 82, 150
heroin chic, 150
hips, 12, 15, 20–1; waist–hip ratio, 49,
    51, 74–6, 80–1, 154
Hispanic ideal female body shape,
    152–3
Hitchcock, Alfred, 150
homosexuality: role of female body

in same-sex relationships, 4; and sexual selection, 69–70
hormone replacement therapy, 35
hormones: and appetite, 109–10, 113, 115; and eating disorders, 125; and fat deposition, 30, 31, 34–6; and fertility, 49–50; and other pubertal changes, 36–7; and sex differentiation, 29; and sexual selection, 68–9, 72–3
hosiery, 174
humour, and sexual selection, 67
hunger, 109
hunting, 25–6
hyperphagia, 118
hypothalamus, 49–50, 68–9, 131

inheritance and genetics: and appetite, 109; and body image, 94–5; and body shape, 43; and breast cancer, 56; and competition, 191; and eating disorders, 128; and sexual selection, 66, 68
insula, 93, 131, 132–3
insulin, 30, 51–2
intelligence: and body symmetry, 71, 80; body's role in, 86–7; and breastfeeding, 23, 39; and dieting, 119; and sexual selection, 62, 66–7, 62, 66–7
internet: and body image, 90; and ideal body shape, 158, 160, 161
Irving, John, 9

jeans, 164, 169–70
Jeffries, Mike, 187
joints, 17, 54
Jolie, Angelina, 82, 156
Jung, Carl, 123

Kayan people, 173
Kennedy, Jacqueline, 150
Khoisan people, 155
kidneys, 54
Knowles, Beyoncé, 82, 153

lactation and breastfeeding: and breast size, 77–8; calories needed for, 38–9; and cancer, 56; and fat deposition, 22; and female athletic abilities, 25–6; and infant intelligence, 23, 39; strategies for meeting the demands, 44
Lagerfeld, Karl, 90
legs, 12, 16, 174–7
Lemburg Castle, 171
leptin, 30, 36, 50, 109–10
Lewis, Matthew Gregory, 27
lice, 165–6
life expectancy, 20
liposuction, 186
lips, 36, 72, 183
liver, 52; disease, 175
Lukyanova, Valeria, 160–1
lungs, 24

magazines, and ideal body shape, 158, 159, 160, 161–2, 162–3
makeover programmes, 162
makeup, 182, 183
Manet, Edouard, 149
manipulation, physical, 86–7
Mansfield, Jayne, 82, 150
marriage, and weight, 114
Mary, Virgin, 149
Masaccio, 164
Mauritania, 152
media, and ideal female body shape, 153, 156–63
men: attitude to makeup, 182; clothing, 168, 172–3, 178–9; and female body concealment, 167; and oestrogens, 35; preferences for female body shape and appearance, 61–83, 152, 156, 175–6, 183, 202–3; societal ideal for body shape, 146; see also sex differences
menopause, and after, 40
menstrual cycle: and appetite, 114–15; and clothing, 180, 198; and fertility, 180
mental health, 71, 77; see also anxiety;

depression
mice, 21, 30, 68
Middle Eastern ideal female body
    shape, 153
mind, relationship to the body, 86–93,
    99–103
mirrors, and body image, 89–90, 100
Monroe, Marilyn, 82, 150, 176, 183
mood: and dieting, 119; and eating,
    108–15; and exercise, 120, 121; *see
    also* emotions
Moore, Demi, 174
movement, 87, 88, 176, 177
muscles, 17, 21, 23, 31, 32
music, and sexual selection, 67
muzzles, 15

nakedness, 164, 167
Navratilova, Martina, 32
necks, and clothing, 173
nerves, sympathetic, 30, 35, 36
noradrenaline, 30
noses, 72
nudity *see* nakedness
Nyström, Lene, 145–6

obesity *see* weight
oesophagus, 125
oestrogens: and appetite, 115; and
    body shape, 35–6, 37; and breast
    symmetry, 80; and facial features,
    183; and the menopause, 40; science
    of, 33–4
osteoporosis, 54
ovaries, 35, 40

Pacific islanders, 155
Paltrow, Gwyneth, 82
pancreas, 51–2
parents: and competition, 191, 192;
    and eating disorders, 134; other-sex
    parents as models of desirability, 66,
    72; and their children's body image,
    96–7
peacocks, 64
pelvises, 12–13, 18–21, 31

personality: and body image, 94;
    depersonalisation disorder, 93;
    and eating disorders, 128–9; and
    reaction to carbohydrate, 111; and
    sexual selection, 67
Philippines, 25
photography: female tricks for
    appearing smaller, 107; and self-
    perception, 89–90
piercings, 185
pituitary gland, 49–50
plastic surgery *see* cosmetic surgery
Playboy magazine, 82, 150–1, 155–6,
    159
pornography, 158–9, 167
prefrontal cortex, 130–1, 132–3
pregnancy: and being overweight, 48;
    clothing, 174; and eating disorders,
    127; and fat deposition, 22; and
    female abdomens, 17; and female
    athletic abilities, 25–6; fetal sex
    development, 28–9; preferred
    intercourse positions during, 76;
    shape of bulge, 76; stance in late
    pregnancy, 20; weight gain during,
    37–8; *see also* childbirth
pregorexia, 127
primates, and female competition,
    191; *see also* chimpanzees
proprioception, 91
puberty: and body image, 96; and
    differences between the sexes, 31–7;
    and eating disorders, 127, 135, 139;
    onset triggers, 45–8, 50
pubic hair *see* hair, pubic and armpit
pubic mounds, 35

rats, 138
reproduction *see* childbirth; fertility;
    lactation and breastfeeding;
    pregnancy
ribcages, 17
right- and left-handedness, 17, 92, 131
Rihanna, 153
Rirgendanwa, Carine, 154
romantic relationships: and appetite,

113–14; and body image, 100–1; and social status, 192; and victims of aggression, 194

Rubens, Peter Paul, vii, 149

sadness, and appetite, 113
schizophrenia, 177
self-awareness, and the body, 84–103
self-esteem, and body image, 101–2
serotonin, 111, 113, 115, 128, 133
sex development, fetal, 28–9
sex differences: aggression, 193; athletic abilities, 24; body-dissatisfaction, 2, 97; body shape, 13–28, 31–44; colour blindness, 86; competition and social status, 190–3, 203–5; diabetes, 52–3; eating disorders, 126–7, 136; eating habits and preferences, 108, 110, 118; exercise, 36; fat metabolism, 22–3, 52; maturation age, 32–3; perceiving body shape change, 88–9; sense of smell, 86; size, 13–15; talking, 193; vertical growth, 36
sexual abuse, and eating disorders, 134
sexual attraction and selection: acceptability of criteria, 202, 207; and body image, 100–1; and body shape, 61–83, 175–6; and clothing, 179–80; and competition, 190–1; and exercise, 120; and skin, 182–3
sexual intercourse: preferred positions in pregnancy, 76; sex-life and body image, 100
sexual jealousy, 184
Shakespeare, William, 189
shoes, high-heeled, 176–7
shoulders, 17, 173
shrews, 21
siblings: identifying by smell, 68; and origins of eating disorders, 139–40
Simpson, Wallis, vii
size, 13–15
skin, 40, 41, 70, 73, 182–4
skulls, 12, 16
smell: and self-perception, 91; sexual

differentiation in sense of, 86; and sexual selection, 68–9
smiles, 67
smoking, and dieting, 118
social roles *see* labour division
social status: and attractiveness, 195–201; and body shape, 154, 199–201; and clothing, 168–9; hierarchy formation, 191–2; importance to women, 190–8, 203–6; male, 204–5; and stress, 191
Some Like It Hot (film), 176
spatial timidity, 106–7
spines *see* backs and spines
sporting abilities *see* athletic abilities
steatopygia, 155
steroids, 34, 36–7; *see also* oestrogens; testosterone
stomach, 125
strength, 17
stress: and eating, 111, 113; and social status, 191
stripes, and clothing, 181–2
Sudanese ideal female body shape, 153
suntans, 183–4
surgery *see* cosmetic surgery
swimming, 120
symmetry, and sexual selection, 70–1, 80

talking, 193–4
tanning *see* suntans
tattoos, 184–5
teeth, 15, 70, 125
television, and ideal female body shape, 153, 157–8, 160, 162
testicles, 29
testosterone, 36–7
thighs: brain's perception of, 88; and breastfeeding, 23, 39; and fat deposition, 21, 22, 23, 31, 35; liposuction, 186; male–female differences, 15; and pregnancy, 39; societal ideal through the ages, 147, 148–9; steatopygia, 155
thoraxes, 17